Regression Analysis and
Linear Models

Regression Analysis and Linear Models

Edited by
Graham Molloy

Larsen & Keller
www.larsen-keller.com

Regression Analysis and Linear Models
Edited by Graham Molloy
ISBN: 978-1-63549-705-2 (Hardback)

Published by Larsen and Keller Education,
5 Penn Plaza,
19th Floor,
New York, NY 10001, USA

Cataloging-in-Publication Data

Regression analysis and linear models / edited by Graham Molloy.
      p. cm.
Includes bibliographical references and index.
ISBN 978-1-63549-705-2
1. Regression analysis. 2. Linear models (Statistics). I. Molloy, Graham.
QA278.2 .R44 2018
519.536--dc23

For more information regarding Larsen and Keller Education and its products, please visit the publisher's website www.larsen-keller.com

Table of Contents

Preface   **VII**

Chapter 1   **Understanding Linear Regression Analysis**   **1**
    a. Linear Regression   1
    b. Regression Analysis   16

Chapter 2   **An Overview of Simple Regression Analysis**   **44**
    a. Simple Linear Regression   44
    b. Variance   77
    c. Covariance   94
    d. Residual Sum of Squares   100
    e. Maximum Likelihood Estimation   106
    f. Analysis of Variance   127
    g. Prediction Interval   145

Chapter 3   **An Integrated Study of Multiple Linear Regression Analysis** **167**
    a. Multiple Linear Regression Analysis   167
    b. Gauss–Markov Theorem   173
    c. Cramér–Rao Bound   184

Chapter 4   **Standardized Coefficient and Statistical Hypothesis Testing** **193**
    a. Standardized Coefficient   193
    b. Statistical Hypothesis Testing   211

Permissions

Index

Preface

As a part of statistical modeling, regression analysis is the process and technique of understanding the relationship between dependent and independent variables. It is used in forecasting and prediction. The models used to understand and build relationships, using the linear predictor function, is known as linear models. The different models that come under this vast field are hierarchical linear model, general linear model, generalized linear model, etc. This book presents the complex subject of regression analysis and linear models, in the most comprehensible and easy to understand language. Such selected concepts that redefine the subject have been presented herein. Those with an interest in the field of regression analysis and linear models would find this textbook helpful.

A detailed account of the significant topics covered in this book is provided below:

Chapter 1- Linear regression attempts to model the relationship between two continuous variables, where there is one scalar variable and one or more dependent variables. Linear regression is generally used in behavioral, biological and social sciences. This chapter has been carefully written to provide an easy understanding of the varied facets of regression analysis and linear models.

Chapter 2- Regression analysis helps in assessing the relationship between variables. It overlaps with fields like machine learning when used for forecasting or prediction. Variance, covariance, residual sum of squares, ANOVA and prediction interval are some of the topics discussed in the chapter. The topics discussed in the chapter are of great importance to broaden the existing knowledge on regression analysis and linear models.

Chapter 3- Multiple linear regression analysis studies the relationship between two or more variables and a dependent variable by including a linear equation so that every independent variable is associated with the value of every dependent variable. The chapter strategically encompasses and incorporates the major components and key concepts of multiple linear regression analysis, providing a complete understanding.

Chapter 4- The estimates that arise due to the standardization of regression analysis, which cause the variables of dependent and independent variables to be 1. Unit normal scaling and unit length scaling are two common methods of standardized regression coefficients. The aspects elucidated in this chapter are of vital importance, and provide a better understanding of regression analysis and linear models.

I would like to make a special mention of my publisher who considered me worthy of this opportunity and also supported me throughout the process. I would also like to thank the editing team at the back-end who extended their help whenever required.

Editor

Understanding Linear Regression Analysis

Linear regression attempts to model the relationship between two continuous variables, where there is one scalar variable and one or more dependent variables. Linear regression is generally used in behavioral, biological and social sciences. This chapter has been carefully written to provide an easy understanding of the varied facets of regression analysis and linear models.

Linear Regression

In statistics, linear regression is an approach for modeling the relationship between a scalar dependent variable y and one or more explanatory variables (or independent variables) denoted X. The case of one explanatory variable is called *simple linear regression*. For more than one explanatory variable, the process is called *multiple linear regression*. (This term is distinct from *multivariate linear regression*, where multiple correlated dependent variables are predicted, rather than a single scalar variable.)

In linear regression, the relationships are modeled using linear predictor functions whose unknown model parameters are estimated from the data. Such models are called *linear models*. Most commonly, the conditional mean of y given the value of X is assumed to be an affine function of X; less commonly, the median or some other quantile of the conditional distribution of y given X is expressed as a linear function of X. Like all forms of regression analysis, linear regression focuses on the conditional probability distribution of y given X, rather than on the joint probability distribution of y and X, which is the domain of multivariate analysis.

Linear regression was the first type of regression analysis to be studied rigorously, and to be used extensively in practical applications. This is because models which depend linearly on their unknown parameters are easier to fit than models which are non-linearly related to their parameters and because the statistical properties of the resulting estimators are easier to determine.

Linear regression has many practical uses. Most applications fall into one of the following two broad categories:

- If the goal is prediction, or forecasting, or error reduction, linear regression can be used to fit a predictive model to an observed data set of y and X values. After developing such a model, if an additional value of X is then given without its accompanying value of y, the fitted model can be used to make a prediction of the value of y.

- Given a variable y and a number of variables X_1, ..., X_p that may be related to y, linear regression analysis can be applied to quantify the strength of the relationship between y and the X_j, to assess which X_j may have no relationship with y at all, and to identify which subsets of the X_j contain redundant information about y.

Linear regression models are often fitted using the least squares approach, but they may also be fitted in other ways, such as by minimizing the "lack of fit" in some other norm (as with least absolute deviations regression), or by minimizing a penalized version of the least squares loss function as in ridge regression (L^2-norm penalty) and lasso (L^1-norm penalty). Conversely, the least squares approach can be used to fit models that are not linear models. Thus, although the terms "least squares" and "linear model" are closely linked, they are not synonymous.

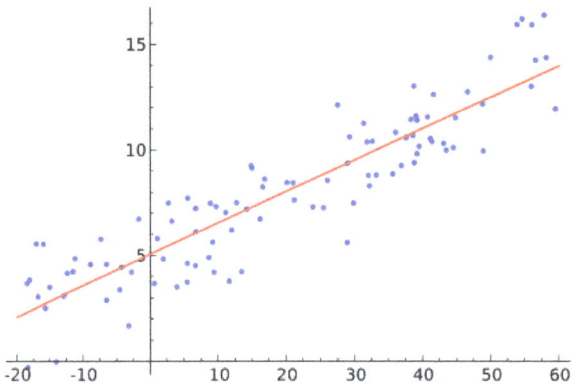

Example of simple linear regression, which has one independent variable.

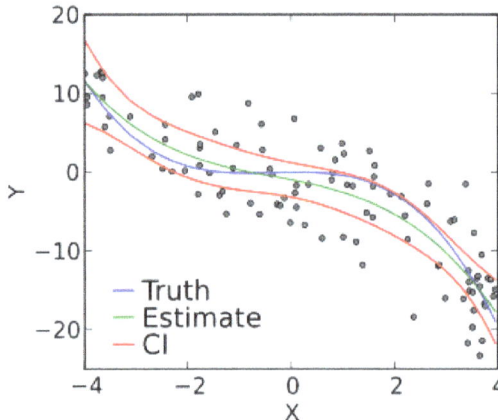

Example of a cubic polynomial regression, which is a type of linear regression.

Given a data set $\{y_i, x_{i1}, \ldots, x_{ip}\}_{i=1}^{n}$ of n statistical units, a linear regression model assumes that the relationship between the dependent variable y_i and the p-vector of regressors $\boldsymbol{x}_i$ is linear. This relationship is modeled through a *disturbance term* or *error variable* ε_i — an unobserved random variable that adds noise to the linear relationship between the dependent variable and regressors. Thus the model takes the form

$$y_i = \beta_0 1 + \beta_1 x_{i1} + \cdots + \beta_p x_{ip} + \varepsilon_i = x_i^\top \beta + \varepsilon_i, \qquad i = 1, \ldots, n,$$

where $^\top$ denotes the transpose, so that $x_i^\top \beta$ is the inner product between vectors x_i and β.

Often these n equations are stacked together and written in vector form as

$$y = X\beta + \varepsilon,$$

where

$$y = \begin{pmatrix} y_1 \\ y_2 \\ \vdots \\ y_n \end{pmatrix},$$

$$X = \begin{pmatrix} x_1^\top \\ x_2^\top \\ \vdots \\ x_n^\top \end{pmatrix} = \begin{pmatrix} 1 & x_{11} & \cdots & x_{1p} \\ 1 & x_{21} & \cdots & x_{2p} \\ \vdots & \vdots & \ddots & \vdots \\ 1 & x_{n1} & \cdots & x_{np} \end{pmatrix},$$

$$\beta = \begin{pmatrix} \beta_0 \\ \beta_1 \\ \beta_2 \\ \vdots \\ \beta_p \end{pmatrix}, \quad \varepsilon = \begin{pmatrix} \varepsilon_1 \\ \varepsilon_2 \\ \vdots \\ \varepsilon_n \end{pmatrix}.$$

Some remarks on terminology and general use:

- y_i is called the *regressand, endogenous variable, response variable, measured variable, criterion variable,* or *dependent variable*. The decision as to which variable in a data set is modeled as the dependent variable and which are modeled as the independent variables may be based on a presumption that the value of one of the variables is caused by, or directly influenced by the other variables. Alternatively, there may be an operational reason to model one of the variables in terms of the others, in which case there need be no presumption of causality.

- $x_{i1}, x_{i2}, \ldots, x_{ip}$ are called *regressors, exogenous variables, explanatory variables, covariates, input variables, predictor variables,* or *independent variables*. The matrix X is sometimes called the design matrix.

  - Usually a constant is included as one of the regressors. For example, we can take $x_{i1} = 1$ for $i = 1, \ldots, n$. The corresponding element of β is called the *intercept*. Many statistical inference procedures for linear models require an intercept to be present, so it is often included even if theoretical considerations suggest that its value should be zero.

  - Sometimes one of the regressors can be a non-linear function of another regressor or of the data, as in polynomial regression and segmented regression. The model remains linear as long as it is linear in the parameter vector β.

  - The regressors x_{ij} may be viewed either as random variables, which we simply observe, or they can be considered as predetermined fixed values which we can choose. Both interpretations may be appropriate in different cases, and they generally lead to the same estimation procedures; however different approaches to asymptotic analysis are used in these two situations.

- β is a *(p+1)*-dimensional *parameter vector*. Where β_0 is the constant (offset) term. In figure Simple linear regression with one variable $\beta_0 = 5$. Its elements are also called *effects*, or *regression coefficients*. Statistical estimation and inference in linear regression focuses on β. The elements of this parameter vector are interpreted as the partial derivatives of the dependent variable with respect to the various independent variables.

- ε_i is called the *error term*, *disturbance term*, or *noise*. This variable captures all other factors which influence the dependent variable y_i other than the regressors x_i. The relationship between the error term and the regressors, for example whether they are correlated, is a crucial step in formulating a linear regression model, as it will determine the method to use for estimation.

Example. Consider a situation where a small ball is being tossed up in the air and then we measure its heights of ascent h_i at various moments in time t_i. Physics tells us that, ignoring the drag, the relationship can be modeled as

$$h_i = \beta_1 t_i + \beta_2 t_i^2 + \varepsilon_i,$$

where β_1 determines the initial velocity of the ball, β_2 is proportional to the standard gravity, and ε_i is due to measurement errors. Linear regression can be used to estimate the values of β_1 and β_2 from the measured data. This model is non-linear in the time variable, but it is linear in the parameters β_1 and β_2; if we take regressors $\mathbf{x}_i = (x_{i1}, x_{i2}) = (t_i, t_i^2)$, the model takes on the standard form

$$h_i = \mathbf{x}_i^\top \beta + \varepsilon_i.$$

Assumptions

Standard linear regression models with standard estimation techniques make a number of assumptions about the predictor variables, the response variables and their relationship. Numerous extensions have been developed that allow each of these assumptions to be relaxed (i.e. reduced to a weaker form), and in some cases eliminated entirely. Some methods are general enough that they can relax multiple assumptions at once, and in other cases this can be achieved by combining different extensions. Generally these extensions make the estimation procedure more complex and time-consuming, and may also require more data in order to produce an equally precise model.

The following are the major assumptions made by standard linear regression models with standard estimation techniques (e.g. ordinary least squares):

- Weak exogeneity. This essentially means that the predictor variables x can be treated as fixed values, rather than random variables. This means, for example, that the predictor variables are assumed to be error-free—that is, not contaminated with measurement errors. Although this assumption is not realistic in many settings, dropping it leads to significantly more difficult errors-in-variables models.

- Linearity. This means that the mean of the response variable is a linear combination of the parameters (regression coefficients) and the predictor variables. Note that this assumption is much less restrictive than it may at first seem. Because the predictor variables are treated as fixed values, linearity is really only a restriction on the parameters. The predictor variables themselves can be arbitrarily transformed, and in fact multiple copies of the same underlying predictor variable can be added, each one transformed differently. This trick is used, for example, in polynomial regression, which uses linear regression to fit the response variable as an arbitrary polynomial function (up to a given rank) of a predictor variable. This makes linear regression an extremely powerful inference method. In fact, models such as polynomial regression are often "too powerful", in that they tend to overfit the data. As a result, some kind of regularization must typically be used to prevent unreasonable solutions coming out of the estimation process. Common examples are ridge regression and lasso regression. Bayesian linear regression can also be used, which by its nature is more or less immune to the problem of overfitting. (In fact, ridge regression and lasso regression can both be viewed as special cases of Bayesian linear regression, with particular types of prior distributions placed on the regression coefficients.)

- Constant variance (a.k.a. homoscedasticity). This means that different response variables have the same variance in their errors, regardless of the values of the predictor variables. In practice this assumption is invalid (i.e. the errors are heteroscedastic) if the response variables can vary over a wide scale. In order to determine for heterogeneous error variance, or when a pattern of residuals violates model assumptions of homoscedasticity (error is equally variable around the 'best-fitting line' for all points of x), it is prudent to look for a "fanning effect" between residual error and predicted values. This is to say there will be a systematic change in the absolute or squared residuals when plotted against the predicting outcome. Error will not be evenly distributed across the regression line. Heteroscedasticity will result in the averaging over of distinguishable variances around the points to get a single variance that is inaccurately representing all the variances of the line. In effect, residuals appear clustered and spread apart on their predicted plots for larger and smaller values for points along the linear regression line, and the mean squared error for the model will be wrong. Typically, for example, a response variable whose mean is large will have a greater variance than one whose mean is small. For example, a given person whose income is predicted to be $100,000 may easily have an actual income of $80,000 or $120,000 (a standard deviation of around $20,000), while another person with a predicted income of $10,000 is unlikely to have the same $20,000 standard deviation, which would imply their actual income would vary anywhere between -$10,000 and $30,000. (In fact, as this shows, in many cases—often the same cases where the assumption of normally distributed errors fails—the variance or standard deviation should be predicted to be proportional to the mean, rather than constant.) Simple linear regression estimation methods give less precise parameter estimates and misleading inferential quantities such as standard errors when substantial heteroscedasticity is present. However, various estimation techniques (e.g. weighted least squares and heteroscedasticity-consistent standard errors) can handle heteroscedasticity in a quite general way. Bayesian linear regression techniques can also be used when the variance is assumed to be a function of the mean. It is also possible in some cases to fix the problem by applying a transformation to the response variable (e.g. fit the logarithm of the response variable using a linear regression model, which implies that the response variable has a log-normal distribution rather than a normal distribution).

- Independence of errors. This assumes that the errors of the response variables are uncorrelated with each other. (Actual statistical independence is a stronger condition than mere lack of correlation and is often not needed, although it can be exploited if it is known to hold.) Some methods (e.g. generalized least squares) are capable of handling correlated errors, although they typically require significantly more data unless some sort of regularization is used to bias the model towards assuming uncorrelated errors. Bayesian linear regression is a general way of handling this issue.

- Lack of multicollinearity in the predictors. For standard least squares estimation methods, the design matrix X must have full column rank p; otherwise, we have a condition known as multicollinearity in the predictor variables. This can be triggered by having two or more perfectly correlated predictor variables (e.g. if the same predictor variable is mistakenly given twice, either without transforming one of the copies or by transforming one of the copies linearly). It can also happen if there is too little data available compared to the number of parameters to be estimated (e.g. fewer data points than regression coefficients). In the case of multicollinearity, the parameter vector β will be non-identifiable—it has no unique solution. At most we will be able to identify some of the parameters, i.e. narrow down its value to some linear subspace of R^p. Methods for fitting linear models with multicollinearity have been developed; some require additional assumptions such as "effect sparsity"—that a large fraction of the effects are exactly zero.

  Note that the more computationally expensive iterated algorithms for parameter estimation, such as those used in generalized linear models, do not suffer from this problem—and in fact it's quite normal when handling categorically valued predictors to introduce a separate indicator variable predictor for each possible category, which inevitably introduces multicollinearity.

Beyond these assumptions, several other statistical properties of the data strongly influence the performance of different estimation methods:

- The statistical relationship between the error terms and the regressors plays an important role in determining whether an estimation procedure has desirable sampling properties such as being unbiased and consistent.

- The arrangement, or probability distribution of the predictor variables x has a major influence on the precision of estimates of β. Sampling and design of experiments are highly developed subfields of statistics that provide guidance for collecting data in such a way to achieve a precise estimate of β.

Interpretation

A fitted linear regression model can be used to identify the relationship between a single predictor variable x_j and the response variable y when all the other predictor variables in the model are "held fixed". Specifically, the interpretation of β_j is the expected change in y for a one-unit change in x_j when the other covariates are held fixed—that is, the expected value of the partial derivative of y with respect to x_j. This is sometimes called the *unique effect* of x_j on y. In contrast, the *marginal effect* of x_j on y can be assessed using a correlation coefficient or simple linear regression model relating x_j to y; this effect is the total derivative of y with respect to x_j.

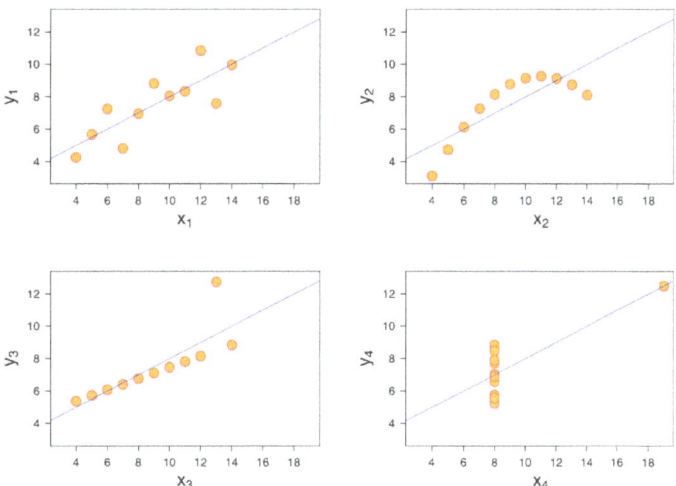

The data sets in the Anscombe's quartet are designed to have
the same linear regression line but are graphically very different.

Care must be taken when interpreting regression results, as some of the regressors may not allow for marginal changes (such as dummy variables, or the intercept term), while others cannot be held fixed (recall the example from the introduction: it would be impossible to "hold t_i fixed" and at the same time change the value of t_i^2).

It is possible that the unique effect can be nearly zero even when the marginal effect is large. This may imply that some other covariate captures all the information in x_j, so that once that variable is in the model, there is no contribution of x_j to the variation in y. Conversely, the unique effect of x_j can be large while its marginal effect is nearly zero. This would happen if the other covariates explained a great deal of the variation of y, but they mainly explain variation in a way that is complementary to what is captured by x_j. In this case, including the other variables in the model reduces the part of the variability of y that is unrelated to x_j, thereby strengthening the apparent relationship with x_j.

The meaning of the expression "held fixed" may depend on how the values of the predictor variables arise. If the experimenter directly sets the values of the predictor variables according to a study design, the comparisons of interest may literally correspond to comparisons among units whose predictor variables have been "held fixed" by the experimenter. Alternatively, the expression "held fixed" can refer to a selection that takes place in the context of data analysis. In this case, we "hold a variable fixed" by restricting our attention to the subsets of the data that happen to have a common value for the given predictor variable. This is the only interpretation of "held fixed" that can be used in an observational study.

The notion of a "unique effect" is appealing when studying a complex system where multiple interrelated components influence the response variable. In some cases, it can literally be interpreted as the causal effect of an intervention that is linked to the value of a predictor variable. However, it has been argued that in many cases multiple regression analysis fails to clarify the relationships between the predictor variables and the response variable when the predictors are correlated with each other and are not assigned following a study design. A commonality analysis may be helpful in disentangling the shared and unique impacts of correlated independent variables.

Extensions

Numerous extensions of linear regression have been developed, which allow some or all of the assumptions underlying the basic model to be relaxed.

Simple and Multiple Regression

The very simplest case of a single scalar predictor variable x and a single scalar response variable y is known as *simple linear regression*. The extension to multiple and/or vector-valued predictor variables (denoted with a capital X) is known as *multiple linear regression*, also known as *multivariable linear regression*. Nearly all real-world regression models involve multiple predictors, and basic descriptions of linear regression are often phrased in terms of the multiple regression model. Note, however, that in these cases the response variable y is still a scalar. Another term *multivariate linear regression* refers to cases where y is a vector, i.e., the same as *general linear regression*.

General Linear Models

The general linear model considers the situation when the response variable Y is not a scalar but a vector. Conditional linearity of $E(y|x) = Bx$ is still assumed, with a matrix B replacing the vector β of the classical linear regression model. Multivariate analogues of Ordinary Least-Squares (OLS) and Generalized Least-Squares (GLS) have been developed. "General linear models" are also called "multivariate linear models". These are not the same as multivariable linear models (also called "multiple linear models").

Heteroscedastic Models

Various models have been created that allow for heteroscedasticity, i.e. the errors for different response variables may have different variances. For example, weighted least squares is a method for estimating linear regression models when the response variables may have different error variances, possibly with correlated errors. Heteroscedasticity-consistent standard errors is an improved method for use with uncorrelated but potentially heteroscedastic errors.

Generalized Linear Models

Generalized linear models (GLMs) are a framework for modeling a response variable y that is bounded or discrete. This is used, for example:

- when modeling positive quantities (e.g. prices or populations) that vary over a large scale—which are better described using a skewed distribution such as the log-normal distribution or Poisson distribution (although GLMs are not used for log-normal data, instead the response variable is simply transformed using the logarithm function);

- when modeling categorical data, such as the choice of a given candidate in an election (which is better described using a Bernoulli distribution/binomial distribution for binary choices, or a categorical distribution/multinomial distribution for multi-way choices), where there are a fixed number of choices that cannot be meaningfully ordered;

- when modeling ordinal data, e.g. ratings on a scale from 0 to 5, where the different outcomes can be ordered but where the quantity itself may not have any absolute meaning (e.g. a rating of 4 may not be "twice as good" in any objective sense as a rating of 2, but simply indicates that it is better than 2 or 3 but not as good as 5).

Generalized linear models allow for an arbitrary *link function g* that relates the mean of the response variable to the predictors, i.e. $E(y) = g(\beta'x)$. The link function is often related to the distribution of the response, and in particular it typically has the effect of transforming between the $(-\infty, \infty)$ range of the linear predictor and the range of the response variable.

Some common examples of GLMs are:

- Poisson regression for count data.

- Logistic regression and probit regression for binary data.

- Multinomial logistic regression and multinomial probit regression for categorical data.

- Ordered probit regression for ordinal data.

Single index models allow some degree of nonlinearity in the relationship between x and y, while preserving the central role of the linear predictor $\beta'x$ as in the classical linear regression model. Under certain conditions, simply applying OLS to data from a single-index model will consistently estimate β up to a proportionality constant.

Hierarchical Linear Models

Hierarchical linear models (or *multilevel regression*) organizes the data into a hierarchy of regressions, for example where A is regressed on B, and B is regressed on C. It is often used where the variables of interest have a natural hierarchical structure such as in educational statistics, where students are nested in classrooms, classrooms are nested in schools, and schools are nested in some administrative grouping, such as a school district. The response variable might be a measure of student achievement such as a test score, and different covariates would be collected at the classroom, school, and school district levels.

Errors-in-variables

Errors-in-variables models (or "measurement error models") extend the traditional linear regression model to allow the predictor variables X to be observed with error. This error causes standard estimators of β to become biased. Generally, the form of bias is an attenuation, meaning that the effects are biased toward zero.

Others

- In Dempster–Shafer theory, or a linear belief function in particular, a linear regression model may be represented as a partially swept matrix, which can be combined with similar matrices representing observations and other assumed normal distributions and state equations. The combination of swept or unswept matrices provides an alternative method for estimating linear regression models.

Estimation Methods

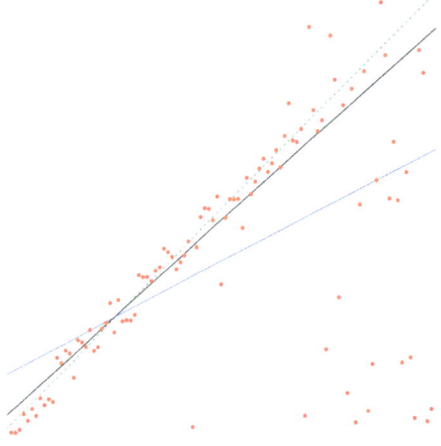

Comparison of the Theil–Sen estimator (black) and simple linear regression (blue) for a set of points with outliers.

A large number of procedures have been developed for parameter estimation and inference in linear regression. These methods differ in computational simplicity of algorithms, presence of a closed-form solution, robustness with respect to heavy-tailed distributions, and theoretical assumptions needed to validate desirable statistical properties such as consistency and asymptotic efficiency.

Some of the more common estimation techniques for linear regression are summarized below.

Least-squares Estimation and Related Techniques

- Ordinary least squares (OLS) is the simplest and thus most common estimator. It is conceptually simple and computationally straightforward. OLS estimates are commonly used to analyze both experimental and observational data.

  The OLS method minimizes the sum of squared residuals, and leads to a closed-form expression for the estimated value of the unknown parameter β:

  $$\hat{\beta} = (\mathbf{X}^\top \mathbf{X})^{-1} \mathbf{X}^\top \mathbf{y} = \left(\sum \mathbf{x}_i \mathbf{x}_i^\top \right)^{-1} \left(\sum \mathbf{x}_i y_i \right).$$

  The estimator is unbiased and consistent if the errors have finite variance and are uncorrelated with the regressors

  $$\mathrm{E}[\mathbf{x}_i \varepsilon_i] = 0.$$

  It is also efficient under the assumption that the errors have finite variance and are homoscedastic, meaning that $\mathrm{E}[\varepsilon_i^2 | \mathbf{x}_i]$ does not depend on i. The condition that the errors are uncorrelated with the regressors will generally be satisfied in an experiment, but in the case of observational data, it is difficult to exclude the possibility of an omitted covariate z that is related to both the observed covariates and the response variable. The existence of such a covariate will generally lead to a correlation between the regressors and the response variable, and hence to an inconsistent estimator of β. The condition of homoscedasticity can fail with either experimental or observational data. If the goal is either inference or predictive modeling, the performance of OLS estimates can be poor if multicollinearity is present, unless the sample size is large.

In simple linear regression, where there is only one regressor (with a constant), the OLS coefficient estimates have a simple form that is closely related to the correlation coefficient between the covariate and the response.

- Generalized least squares (GLS) is an extension of the OLS method, that allows efficient estimation of β when either heteroscedasticity, or correlations, or both are present among the error terms of the model, as long as the form of heteroscedasticity and correlation is known independently of the data. To handle heteroscedasticity when the error terms are uncorrelated with each other, GLS minimizes a weighted analogue to the sum of squared residuals from OLS regression, where the weight for the ith case is inversely proportional to var(ε_i). This special case of GLS is called "weighted least squares". The GLS solution to estimation problem is

$$\hat{\beta} = (\mathbf{X}^{\top}\Omega^{-1}\mathbf{X})^{-1}\mathbf{X}^{\top}\Omega^{-1}\mathbf{y},$$

  where Ω is the covariance matrix of the errors. GLS can be viewed as applying a linear transformation to the data so that the assumptions of OLS are met for the transformed data. For GLS to be applied, the covariance structure of the errors must be known up to a multiplicative constant.

- Percentage least squares focuses on reducing percentage errors, which is useful in the field of forecasting or time series analysis. It is also useful in situations where the dependent variable has a wide range without constant variance, as here the larger residuals at the upper end of the range would dominate if OLS were used. When the percentage or relative error is normally distributed, least squares percentage regression provides maximum likelihood estimates. Percentage regression is linked to a multiplicative error model, whereas OLS is linked to models containing an additive error term.

- Iteratively reweighted least squares (IRLS) is used when heteroscedasticity, or correlations, or both are present among the error terms of the model, but where little is known about the covariance structure of the errors independently of the data. In the first iteration, OLS, or GLS with a provisional covariance structure is carried out, and the residuals are obtained from the fit. Based on the residuals, an improved estimate of the covariance structure of the errors can usually be obtained. A subsequent GLS iteration is then performed using this estimate of the error structure to define the weights. The process can be iterated to convergence, but in many cases, only one iteration is sufficient to achieve an efficient estimate of β.

- Instrumental variables regression (IV) can be performed when the regressors are correlated with the errors. In this case, we need the existence of some auxiliary *instrumental variables* z_i such that $E[z_i \varepsilon_i] = 0$. If Z is the matrix of instruments, then the estimator can be given in closed form as

$$\hat{\beta} = (\mathbf{X}^{\top}\mathbf{Z}(\mathbf{Z}^{\top}\mathbf{Z})^{-1}\mathbf{Z}^{\top}\mathbf{X})^{-1}\mathbf{X}^{\top}\mathbf{Z}(\mathbf{Z}^{\top}\mathbf{Z})^{-1}\mathbf{Z}^{\top}\mathbf{y}.$$

- Optimal instruments regression is an extension of classical IV regression to the situation where $E[\varepsilon_i \mid z_i] = 0$.

- Total least squares (TLS) is an approach to least squares estimation of the linear regression model that treats the covariates and response variable in a more geometrically symmetric manner than OLS. It is one approach to handling the "errors in variables" problem, and is also sometimes used even when the covariates are assumed to be error-free.

Maximum-likelihood Estimation and Related Techniques

- Maximum likelihood estimation can be performed when the distribution of the error terms is known to belong to a certain parametric family f_θ of probability distributions. When f_θ is a normal distribution with zero mean and variance θ, the resulting estimate is identical to the OLS estimate. GLS estimates are maximum likelihood estimates when ε follows a multivariate normal distribution with a known covariance matrix.

- Ridge regression, and other forms of penalized estimation such as Lasso regression, deliberately introduce bias into the estimation of β in order to reduce the variability of the estimate. The resulting estimators generally have lower mean squared error than the OLS estimates, particularly when multicollinearity is present or when overfitting is a problem. They are generally used when the goal is to predict the value of the response variable y for values of the predictors x that have not yet been observed. These methods are not as commonly used when the goal is inference, since it is difficult to account for the bias.

- Least absolute deviation (LAD) regression is a robust estimation technique in that it is less sensitive to the presence of outliers than OLS (but is less efficient than OLS when no outliers are present). It is equivalent to maximum likelihood estimation under a Laplace distribution model for ε.

- Adaptive estimation. If we assume that error terms are independent from the regressors $\varepsilon_i \perp \mathbf{x}_i$, the optimal estimator is the 2-step MLE, where the first step is used to non-parametrically estimate the distribution of the error term.

Other Estimation Techniques

- Bayesian linear regression applies the framework of Bayesian statistics to linear regression. In particular, the regression coefficients β are assumed to be random variables with a specified prior distribution. The prior distribution can bias the solutions for the regression coefficients, in a way similar to (but more general than) ridge regression or lasso regression. In addition, the Bayesian estimation process produces not a single point estimate for the "best" values of the regression coefficients but an entire posterior distribution, completely describing the uncertainty surrounding the quantity. This can be used to estimate the "best" coefficients using the mean, mode, median, any quantile, or any other function of the posterior distribution.

- Quantile regression focuses on the conditional quantiles of y given X rather than the conditional mean of y given X. Linear quantile regression models a particular conditional quantile, for example the conditional median, as a linear function $\beta^T x$ of the predictors.

- Mixed models are widely used to analyze linear regression relationships involving dependent data when the dependencies have a known structure. Common applications of mixed models include analysis of data involving repeated measurements, such as longitudinal data, or data obtained from cluster sampling. They are generally fit as parametric models, using maximum likelihood or Bayesian estimation. In the case where the errors are modeled as normal random variables, there is a close connection between mixed models and generalized least squares. Fixed effects estimation is an alternative approach to analyzing this type of data.

- Principal component regression (PCR) is used when the number of predictor variables is large, or when strong correlations exist among the predictor variables. This two-stage procedure first reduces the predictor variables using principal component analysis then uses the reduced variables in an OLS regression fit. While it often works well in practice, there is no general theoretical reason that the most informative linear function of the predictor variables should lie among the dominant principal components of the multivariate distribution of the predictor variables. The partial least squares regression is the extension of the PCR method which does not suffer from the mentioned deficiency.

- Least-angle regression is an estimation procedure for linear regression models that was developed to handle high-dimensional covariate vectors, potentially with more covariates than observations.

- The Theil–Sen estimator is a simple robust estimation technique that chooses the slope of the fit line to be the median of the slopes of the lines through pairs of sample points. It has similar statistical efficiency properties to simple linear regression but is much less sensitive to outliers.

- Other robust estimation techniques, including the α-trimmed mean approach, and L-, M-, S-, and R-estimators have been introduced.

Further Discussion

In statistics and numerical analysis, the problem of numerical methods for linear least squares is an important one because linear regression models are one of the most important types of model, both as formal statistical models and for exploration of data sets. The majority of statistical computer packages contain facilities for regression analysis that make use of linear least squares computations. Hence it is appropriate that considerable effort has been devoted to the task of ensuring that these computations are undertaken efficiently and with due regard to numerical precision.

Individual statistical analyses are seldom undertaken in isolation, but rather are part of a sequence of investigatory steps. Some of the topics involved in considering numerical methods for linear least squares relate to this point. Thus important topics can be:

- Computations where a number of similar, and often nested, models are considered for the same data set. That is, where models with the same dependent variable but different sets of independent variables are to be considered, for essentially the same set of data points.

- Computations for analyses that occur in a sequence, as the number of data points increases.

- Special considerations for very extensive data sets.

Fitting of linear models by least squares often, but not always, arises in the context of statistical analysis. It can therefore be important that considerations of computational efficiency for such problems extend to all of the auxiliary quantities required for such analyses, and are not restricted to the formal solution of the linear least squares problem.

Matrix calculations, like any others, are affected by rounding errors. An early summary of these effects, regarding the choice of computational methods for matrix inversion, was provided by Wilkinson.

Using Linear Algebra

It follows that one can find a "best" approximation of another function by minimizing the area between two functions, a continuous function f on $[a,b]$ and a function $g \in W$ where W is a subspace of $C[a,b]$:

$$\text{Area} = \int_a^b |f(x) - g(x)| dx,$$

all within the subspace W. Due to the frequent difficulty of evaluating integrands involving absolute value, one can instead define

$$\int_a^b [f(x) - g(x)]^2 dx$$

as an adequate criterion for obtaining the least squares approximation, function g, of f with respect to the inner product space W.

As such, $\| f - g \|^2$ or, equivalently, $\| f - g \|$, can thus be written in vector form:

$$\int_a^b [f(x) - g(x)]^2 dx = \langle f - g, f - g \rangle = \| f - g \|^2$$

In other words, the least squares approximation of f is the function $g \in$ subspace W closest to f in terms of the inner product $\langle f, g \rangle$. Furthermore, this can be applied with a theorem:

> Let f be continuous on $[a,b]$, and let W be a finite-dimensional subspace of $C[a,b]$. The least squares approximating function of f with respect to W is given by
>
> $$g = \langle f, \vec{w}_1 \rangle \vec{w}_1 + \langle f, \vec{w}_2 \rangle \vec{w}_2 + \cdots + \langle f, \vec{w}_n \rangle \vec{w}_n,$$
>
> where $B = \{\vec{w}_1, \vec{w}_2, \ldots, \vec{w}_n\}$ is an orthonormal basis for W.

Applications of Linear Regression

Linear regression is widely used in biological, behavioral and social sciences to describe possible relationships between variables. It ranks as one of the most important tools used in these disciplines.

Trend Line

A trend line represents a trend, the long-term movement in time series data after other components have been accounted for. It tells whether a particular data set (say GDP, oil prices or stock prices) have increased or decreased over the period of time. A trend line could simply be drawn by eye through a set of data points, but more properly their position and slope is calculated using statistical techniques like linear regression. Trend lines typically are straight lines, although some variations use higher degree polynomials depending on the degree of curvature desired in the line.

Trend lines are sometimes used in business analytics to show changes in data over time. This has the advantage of being simple. Trend lines are often used to argue that a particular action or event (such as training, or an advertising campaign) caused observed changes at a point in time. This is a simple technique, and does not require a control group, experimental design, or a sophisticated analysis technique. However, it suffers from a lack of scientific validity in cases where other potential changes can affect the data.

Epidemiology

Early evidence relating tobacco smoking to mortality and morbidity came from observational studies employing regression analysis. In order to reduce spurious correlations when analyzing observational data, researchers usually include several variables in their regression models in addition to the variable of primary interest. For example, suppose we have a regression model in which cigarette smoking is the independent variable of interest, and the dependent variable is lifespan measured in years. Researchers might include socio-economic status as an additional independent variable, to ensure that any observed effect of smoking on lifespan is not due to some effect of education or income. However, it is never possible to include all possible confounding variables in an empirical analysis. For example, a hypothetical gene might increase mortality and also cause people to smoke more. For this reason, randomized controlled trials are often able to generate more compelling evidence of causal relationships than can be obtained using regression analyses of observational data. When controlled experiments are not feasible, variants of regression analysis such as instrumental variables regression may be used to attempt to estimate causal relationships from observational data.

Finance

The capital asset pricing model uses linear regression as well as the concept of beta for analyzing and quantifying the systematic risk of an investment. This comes directly from the beta coefficient of the linear regression model that relates the return on the investment to the return on all risky assets.

Economics

Linear regression is the predominant empirical tool in economics. For example, it is used to predict consumption spending, fixed investment spending, inventory investment, purchases of a country's exports, spending on imports, the demand to hold liquid assets, labor demand, and labor supply.

Environmental Science

Linear regression finds application in a wide range of environmental science applications. In Canada, the Environmental Effects Monitoring Program uses statistical analyses on fish and benthic surveys to measure the effects of pulp mill or metal mine effluent on the aquatic ecosystem.

Regression Analysis

In statistical modeling, regression analysis is a statistical process for estimating the relationships among variables. It includes many techniques for modeling and analyzing several variables, when the focus is on the relationship between a dependent variable and one or more independent variables (or 'predictors'). More specifically, regression analysis helps one understand how the typical value of the dependent variable (or 'criterion variable') changes when any one of the independent variables is varied, while the other independent variables are held fixed. Most commonly, regression analysis estimates the conditional expectation of the dependent variable given the independent variables – that is, the average value of the dependent variable when the independent variables are fixed. Less commonly, the focus is on a quantile, or other location parameter of the conditional distribution of the dependent variable given the independent variables. In all cases, the estimation target is a function of the independent variables called the regression function. In regression analysis, it is also of interest to characterize the variation of the dependent variable around the regression function which can be described by a probability distribution. A related but distinct approach is necessary condition analysis (NCA), which estimates the maximum (rather than average) value of the dependent variable for a given value of the independent variable (ceiling line rather than central line) in order to identify what value of the independent variable is necessary but not sufficient for a given value of the dependent variable.

Regression analysis is widely used for prediction and forecasting, where its use has substantial overlap with the field of machine learning. Regression analysis is also used to understand which among the independent variables are related to the dependent variable, and to explore the forms of these relationships. In restricted circumstances, regression analysis can be used to infer causal relationships between the independent and dependent variables. However this can lead to illusions or false relationships, so caution is advisable; for example, correlation does not imply causation.

Many techniques for carrying out regression analysis have been developed. Familiar methods such as linear regression and ordinary least squares regression are parametric, in that the regression function is defined in terms of a finite number of unknown parameters that are estimated from the data. Nonparametric regression refers to techniques that allow the regression function to lie in a specified set of functions, which may be infinite-dimensional.

The performance of regression analysis methods in practice depends on the form of the data generating process, and how it relates to the regression approach being used. Since the true form of the data-generating process is generally not known, regression analysis often depends to some extent on making assumptions about this process. These assumptions are sometimes testable if a sufficient quantity of data is available. Regression models for prediction are often useful even when the assumptions are moderately violated, although they may not perform optimally. However, in many applications, especially with small effects or questions of causality based on observational data, regression methods can give misleading results.

In a narrower sense, regression may refer specifically to the estimation of continuous response variables, as opposed to the discrete response variables used in classification. The case of a continuous output variable may be more specifically referred to as metric regression to distinguish it from related problems.

History

The earliest form of regression was the method of least squares, which was published by Legendre in 1805, and by Gauss in 1809. Legendre and Gauss both applied the method to the problem of determining, from astronomical observations, the orbits of bodies about the Sun (mostly comets, but also later the then newly discovered minor planets). Gauss published a further development of the theory of least squares in 1821, including a version of the Gauss–Markov theorem.

The term "regression" was coined by Francis Galton in the nineteenth century to describe a biological phenomenon. The phenomenon was that the heights of descendants of tall ancestors tend to regress down towards a normal average (a phenomenon also known as regression toward the mean). For Galton, regression had only this biological meaning, but his work was later extended by Udny Yule and Karl Pearson to a more general statistical context. In the work of Yule and Pearson, the joint distribution of the response and explanatory variables is assumed to be Gaussian. This assumption was weakened by R.A. Fisher in his works of 1922 and 1925. Fisher assumed that the conditional distribution of the response variable is Gaussian, but the joint distribution need not be. In this respect, Fisher's assumption is closer to Gauss's formulation of 1821.

In the 1950s and 1960s, economists used electromechanical desk calculators to calculate regressions. Before 1970, it sometimes took up to 24 hours to receive the result from one regression.

Regression methods continue to be an area of active research. In recent decades, new methods have been developed for robust regression, regression involving correlated responses such as time series and growth curves, regression in which the predictor (independent variable) or response variables are curves, images, graphs, or other complex data objects, regression methods accommodating various types of missing data, nonparametric regression, Bayesian methods for regression, regression in which the predictor variables are measured with error, regression with more predictor variables than observations, and causal inference with regression.

Regression Models

Regression models involve the following variables:

- The unknown parameters, denoted as β, which may represent a scalar or a vector.

- The independent variables, X.

- The dependent variable, Y.

In various fields of application, different terminologies are used in place of dependent and independent variables.

A regression model relates Y to a function of X and β.

$$Y \approx f(X, \beta)$$

The approximation is usually formalized as $E(Y \mid X) = f(X, \beta)$. To carry out regression analysis, the form of the function f must be specified. Sometimes the form of this function is based on knowledge about the relationship between Y and X that does not rely on the data. If no such knowledge is available, a flexible or convenient form for f is chosen.

Assume now that the vector of unknown parameters β is of length k. In order to perform a regression analysis the user must provide information about the dependent variable Y:

- If N data points of the form (Y, X) are observed, where $N < k$, most classical approaches to regression analysis cannot be performed: since the system of equations defining the regression model is underdetermined, there are not enough data to recover β.

- If exactly $N = k$ data points are observed, and the function f is linear, the equations $Y = f(X, \beta)$ can be solved exactly rather than approximately. This reduces to solving a set of N equations with N unknowns (the elements of β), which has a unique solution as long as the X are linearly independent. If f is nonlinear, a solution may not exist, or many solutions may exist.

- The most common situation is where $N > k$ data points are observed. In this case, there is enough information in the data to estimate a unique value for β that best fits the data in some sense, and the regression model when applied to the data can be viewed as an overdetermined system in β.

In the last case, the regression analysis provides the tools for:

1. Finding a solution for unknown parameters β that will, for example, minimize the distance between the measured and predicted values of the dependent variable Y (also known as method of least squares).

2. Under certain statistical assumptions, the regression analysis uses the surplus of information to provide statistical information about the unknown parameters β and predicted values of the dependent variable Y.

Necessary Number of Independent Measurements

Consider a regression model which has three unknown parameters, β_0, β_1, and β_2. Suppose an experimenter performs 10 measurements all at exactly the same value of independent variable vector X (which contains the independent variables X_1, X_2, and X_3). In this case, regression analysis fails to give a unique set of estimated values for the three unknown parameters; the experimenter did not provide enough information. The best one can do is to estimate the average value and the standard deviation of the dependent variable Y. Similarly, measuring at two different values of X would give enough data for a regression with two unknowns, but not for three or more unknowns.

If the experimenter had performed measurements at three different values of the independent variable vector X, then regression analysis would provide a unique set of estimates for the three unknown parameters in β.

In the case of general linear regression, the above statement is equivalent to the requirement that the matrix X^TX is invertible.

Statistical Assumptions

When the number of measurements, N, is larger than the number of unknown parameters, k, and the measurement errors ε_i are normally distributed then *the excess of information* contained in $(N - k)$ measurements is used to make statistical predictions about the unknown parameters. This excess of information is referred to as the degrees of freedom of the regression.

Underlying Assumptions

Classical assumptions for regression analysis include:

- The sample is representative of the population for the inference prediction.

- The error is a random variable with a mean of zero conditional on the explanatory variables.

- The independent variables are measured with no error. (Note: If this is not so, modeling may be done instead using errors-in-variables model techniques).

- The independent variables (predictors) are linearly independent, i.e. it is not possible to express any predictor as a linear combination of the others.

- The errors are uncorrelated, that is, the variance–covariance matrix of the errors is diagonal and each non-zero element is the variance of the error.

- The variance of the error is constant across observations (homoscedasticity). If not, weighted least squares or other methods might instead be used.

These are sufficient conditions for the least-squares estimator to possess desirable properties; in particular, these assumptions imply that the parameter estimates will be unbiased, consistent, and efficient in the class of linear unbiased estimators. It is important to note that actual data rarely satisfies the assumptions. That is, the method is used even though the assumptions are not true. Variation from the assumptions can sometimes be used as a measure of how far the model is from being useful. Many of these assumptions may be relaxed in more advanced treatments. Reports of statistical analyses usually include analyses of tests on the sample data and methodology for the fit and usefulness of the model.

Assumptions include the geometrical support of the variables. Independent and dependent variables often refer to values measured at point locations. There may be spatial trends and spatial autocorrelation in the variables that violate statistical assumptions of regression. Geographic weighted regression is one technique to deal with such data. Also, variables may include values aggregated by areas. With aggregated data the modifiable areal unit problem can cause extreme variation in regression parameters. When analyzing data aggregated by political boundaries, postal codes or census areas results may be very distinct with a different choice of units.

Linear Regression

In linear regression, the model specification is that the dependent variable, y_i is a linear combination of the *parameters* (but need not be linear in the *independent variables*). For example, in simple linear regression for modeling n data points there is one independent variable: x_i, and two parameters, β_0 and β_1:

straight line:

$$y_i = \beta_0 + \beta_1 x_i + \varepsilon_i, \quad i = 1, \ldots, n.$$

In multiple linear regression, there are several independent variables or functions of independent variables.

Adding a term in x_i^2 to the preceding regression gives:

parabola:

$$y_i = \beta_0 + \beta_1 x_i + \beta_2 x_i^2 + \varepsilon_i, i = 1, \ldots, n.$$

This is still linear regression; although the expression on the right hand side is quadratic in the independent variable x_i, it is linear in the parameters β_0, β_1 and β_2.

In both cases, ε_i is an error term and the subscript i indexes a particular observation.

Returning our attention to the straight line case: Given a random sample from the population, we estimate the population parameters and obtain the sample linear regression model:

$$\widehat{y_i} = \hat{\beta}_0 + \hat{\beta}_1 x_i$$

The residual, $e_i = y_i - \hat{y}_i$, is the difference between the value of the dependent variable predicted by the model, $\widehat{y_i}$, and the true value of the dependent variable, y_i. One method of estimation is ordinary least squares. This method obtains parameter estimates that minimize the sum of squared residuals, SSE, also sometimes denoted RSS:

$$SSE = \sum_{i=1}^{n} e_i^2.$$

Minimization of this function results in a set of normal equations, a set of simultaneous linear equations in the parameters, which are solved to yield the parameter estimators, $\hat{\beta}_0, \hat{\beta}_1$.

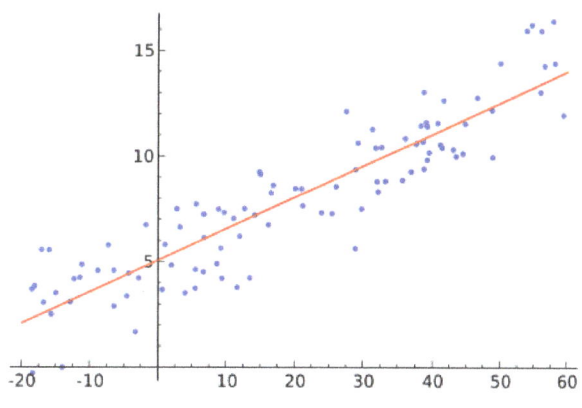

Illustration of linear regression on a data set.

In the case of simple regression, the formulas for the least squares estimates are

$$\widehat{\beta}_1 = \frac{\sum (x_i - \overline{x})(y_i - \overline{y})}{\sum (x_i - \overline{x})^2} \text{ and } \widehat{\beta}_0 = \overline{y} - \widehat{\beta}_1 \overline{x}$$

where $\overline{x}$ is the mean (average) of the x values and $\overline{y}$ is the mean of the y values.

Under the assumption that the population error term has a constant variance, the estimate of that variance is given by:

$$\hat{\sigma}_\varepsilon^2 = \frac{SSE}{n-2}.$$

This is called the mean square error (MSE) of the regression. The denominator is the sample size reduced by the number of model parameters estimated from the same data, (n-p) for p regressors or (n-p-1) if an intercept is used. In this case, p=1 so the denominator is n-2.

The standard errors of the parameter estimates are given by

$$\hat{\sigma}_{\beta_1} = \hat{\sigma}_\varepsilon \sqrt{\frac{1}{\sum (x_i - \overline{x})^2}}.$$

$$\hat{\sigma}_{\beta_0} = \hat{\sigma}_\varepsilon \sqrt{\frac{1}{n} + \frac{\overline{x}^2}{\sum (x_i - \overline{x})^2}}$$

Under the further assumption that the population error term is normally distributed, the researcher can use these estimated standard errors to create confidence intervals and conduct hypothesis tests about the population parameters.

General Linear Model

In the more general multiple regression model, there are p independent variables:

$$y_i = \beta_1 x_{i1} + \beta_2 x_{i2} + \cdots + \beta_p x_{ip} + \varepsilon_i,$$

where x_{ij} is the ith observation on the jth independent variable. If the first independent variable takes the value 1 for all i, $x_{i1} = 1$, then β_1 is called the regression intercept.

The least squares parameter estimates are obtained from p normal equations. The residual can be written as

$$\varepsilon_i = y_i - \hat{\beta}_1 x_{i1} - \cdots - \hat{\beta}_p x_{ip}.$$

The normal equations are

$$\sum_{i=1}^{n}\sum_{k=1}^{p} X_{ij} X_{ik} \hat{\beta}_k = \sum_{i=1}^{n} X_{ij} y_i, \; j = 1,\ldots,p.$$

In matrix notation, the normal equations are written as

$$(X^\top X)\hat{\beta} = X^\top Y,$$

where the ij element of X is x_{ij}, the i element of the column vector Y is y_i, and the j element of $\hat{\beta}$ is $\hat{\beta}_j$. Thus X is $n \times p$, Y is $n \times 1$, and $\hat{\beta}$ is $p \times 1$. The solution is

$$\hat{\beta} = (\mathbf{X}^\top \mathbf{X})^{-1} \mathbf{X}^\top \mathbf{Y}.$$

Diagnostics

Once a regression model has been constructed, it may be important to confirm the goodness of fit of the model and the statistical significance of the estimated parameters. Commonly used checks of goodness of fit include the R-squared, analyses of the pattern of residuals and hypothesis testing. Statistical significance can be checked by an F-test of the overall fit, followed by t-tests of individual parameters.

Interpretations of these diagnostic tests rest heavily on the model assumptions. Although examination of the residuals can be used to invalidate a model, the results of a t-test or F-test are sometimes more difficult to interpret if the model's assumptions are violated. For example, if the error term does not have a normal distribution, in small samples the estimated parameters will not follow normal distributions and complicate inference. With relatively large samples, however, a central limit theorem can be invoked such that hypothesis testing may proceed using asymptotic approximations.

"Limited Dependent" Variables

The phrase "limited dependent" is used in econometric statistics for categorical and constrained variables.

The response variable may be non-continuous ("limited" to lie on some subset of the real line). For binary (zero or one) variables, if analysis proceeds with least-squares linear regression, the model is called the linear probability model. Nonlinear models for binary dependent variables include the probit and logit model. The multivariate probit model is a standard method of estimating a joint relationship between several binary dependent variables and some independent variables. For categorical variables with more than two values there is the multinomial logit. For ordinal variables with more than two values, there are the ordered logit and ordered probit models. Censored regression models may be used when the dependent variable is only sometimes observed, and Heckman correction type models may be used when the sample is not randomly selected from the population of interest. An alternative to such procedures is linear regression based on polychoric correlation (or polyserial correlations) between the categorical variables. Such procedures differ in the assumptions made about the distribution of the variables in the population. If the variable is positive with low values and represents the repetition of the occurrence of an event, then count models like the Poisson regression or the negative binomial model may be used instead.

Interpolation and Extrapolation

Regression models predict a value of the Y variable given known values of the X variables. Prediction *within* the range of values in the dataset used for model-fitting is known informally as interpolation. Prediction *outside* this range of the data is known as extrapolation. Performing extrapolation relies strongly on the regression assumptions. The further the extrapolation goes outside the data, the more room there is for the model to fail due to differences between the assumptions and the sample data or the true values.

It is generally advised that when performing extrapolation, one should accompany the estimated value of the dependent variable with a prediction interval that represents the uncertainty. Such intervals tend to expand rapidly as the values of the independent variable(s) moved outside the range covered by the observed data.

For such reasons and others, some tend to say that it might be unwise to undertake extrapolation.

However, this does not cover the full set of modelling errors that may be being made: in particular, the assumption of a particular form for the relation between Y and X. A properly conducted regression analysis will include an assessment of how well the assumed form is matched by the observed data, but it can only do so within the range of values of the independent variables actually available. This means that any extrapolation is particularly reliant on the assumptions being made about the structural form of the regression relationship. Best-practice advice here is that a linear-in-variables and linear-in-parameters relationship should not be chosen simply for computational convenience, but that all available knowledge should be deployed in constructing a regression model. If this knowledge includes the fact that the dependent variable cannot go outside a certain range of values, this can be made use of in selecting the model – even if the observed dataset has no values particularly near such bounds. The implications of this step of choosing an appropriate functional form for the regression can be great when extrapolation is considered. At a minimum, it can ensure that any extrapolation arising from a fitted model is "realistic" (or in accord with what is known).

Nonlinear Regression

When the model function is not linear in the parameters, the sum of squares must be minimized by an iterative procedure. This introduces many complications which are summarized in Differences between linear and non-linear least squares.

Power and Sample Size Calculations

There are no generally agreed methods for relating the number of observations versus the number of independent variables in the model. One rule of thumb suggested by Good and Hardin is $N = m^n$, where N is the sample size, n is the number of independent variables and m is the number of observations needed to reach the desired precision if the model had only one independent variable. For example, a researcher is building a linear regression model using a dataset that contains 1000 patients (N). If the researcher decides that five observations are needed to precisely define a straight line $((m)$, then the maximum number of independent variables the model can support is 4, because

$$\frac{\log 1000}{\log 5} = 4.29.$$

Other Methods

Although the parameters of a regression model are usually estimated using the method of least squares, other methods which have been used include:

- Bayesian methods, e.g. Bayesian linear regression

- Percentage regression, for situations where reducing *percentage* errors is deemed more appropriate.

- Least absolute deviations, which is more robust in the presence of outliers, leading to quantile regression

- Nonparametric regression, requires a large number of observations and is computationally intensive

- Distance metric learning, which is learned by the search of a meaningful distance metric in a given input space.

Software

All major statistical software packages perform least squares regression analysis and inference. Simple linear regression and multiple regression using least squares can be done in some spreadsheet applications and on some calculators. While many statistical software packages can perform various types of nonparametric and robust regression, these methods are less standardized; different software packages implement different methods, and a method with a given name may be implemented differently in different packages. Specialized regression software has been developed for use in fields such as survey analysis and neuroimaging.

Regression Validation

In statistics, regression validation is the process of deciding whether the numerical results quantifying hypothesized relationships between variables, obtained from regression analysis, are acceptable as descriptions of the data. The validation process can involve analyzing the goodness of fit of the regression, analyzing whether the regression residuals are random, and checking whether the model's predictive performance deteriorates substantially when applied to data that were not used in model estimation.

Validation using R$_2$

An R^2 (coefficient of determination) close to one does not guarantee that the model fits the data well, because as Anscombe's quartet shows, a high R^2 can occur in the presence of misspecification of the functional form of a relationship or in the presence of outliers that distort the true relationship.

One problem with the R^2 as a measure of model validity is that it can always be increased by adding more variables into the model, except in the unlikely event that the additional variables are exactly uncorrelated with the dependent variable in the data sample being used.

Analysis of Residuals

The residuals from a fitted model are the differences between the responses observed at each combination values of the explanatory variables and the corresponding prediction of the response computed using the regression function. Mathematically, the definition of the residual for the ith observation in the data set is written

$$e_i = y_i - f(x_i; \hat{\beta}),$$

with y_i denoting the ith response in the data set and x_i the vector of explanatory variables, each set at the corresponding values found in the ith observation in the data set.

If the model fit to the data were correct, the residuals would approximate the random errors that make the relationship between the explanatory variables and the response variable a statistical relationship. Therefore, if the residuals appear to behave randomly, it suggests that the model fits the data well. On the other hand, if non-random structure is evident in the residuals, it is a clear sign that the model fits the data poorly.

Graphical Analysis of Residuals

A basic, though not quantitatively precise, way to check for problems that render a model inadequate is to conduct a visual examination of the residuals (the mispredictions of the data used in quantifying the model) to look for obvious deviations from randomness. If a visual examination suggests, for example, the possible presence of heteroskedasticity (a relationship between the variance of the model errors and the size of an independent variable's observations), then statistical tests can be performed to confirm or reject this hunch; if it is confirmed, different modeling procedures are called for.

Different types of plots of the residuals from a fitted model provide information on the adequacy of different aspects of the model.

1. sufficiency of the functional part of the model: scatter plots of residuals versus predictors

2. non-constant variation across the data: scatter plots of residuals versus predictors; for data collected over time, also plots of residuals against time

3. drift in the errors (data collected over time): run charts of the response and errors versus time

4. independence of errors: lag plot

5. normality of errors: histogram and normal probability plot

Graphical methods have an advantage over numerical methods for model validation because they readily illustrate a broad range of complex aspects of the relationship between the model and the data.

Quantitative Analysis of Residuals

Numerical methods also play an important role in model validation. For example, the lack-of-fit test for assessing the correctness of the functional part of the model can aid in interpreting a borderline residual plot. One common situation when numerical validation methods take precedence over graphical methods is when the number of parameters being estimated is relatively close to the size of the data set. In this situation residual plots are often difficult to interpret due to constraints on the residuals imposed by the estimation of the unknown parameters. One area in which this typically happens is in optimization applications using designed experiments. Logistic regression with binary data is another area in which graphical residual analysis can be difficult.

Serial correlation of the residuals can indicate model misspecification, and can be checked for with the Durbin–Watson statistic. The problem of heteroskedasticity can be checked for in any of several ways.

Out-of-sample Evaluation

Cross-validation is the process of assessing how the results of a statistical analysis will generalize to an independent data set. If the model has been estimated over some, but not all, of the available data, then the model using the estimated parameters can be used to predict the held-back data. If, for example, the out-of-sample mean squared error, also known as the mean squared prediction error, is substantially higher than the in-sample mean square error, this is a sign of deficiency in the model.

Least Squares

The method of least squares is a standard approach in regression analysis to the approximate solution of overdetermined systems, i.e., sets of equations in which there are more equations than unknowns. "Least squares" means that the overall solution minimizes the sum of the squares of the residuals made in the results of every single equation.

The most important application is in data fitting. The best fit in the least-squares sense minimizes *the sum of squared residuals* (a residual being: the difference between an observed value, and the fitted value provided by a model). When the problem has substantial uncertainties in the independent variable (the x variable), then simple regression and least squares methods have problems; in such cases, the methodology required for fitting errors-in-variables models may be considered instead of that for least squares.

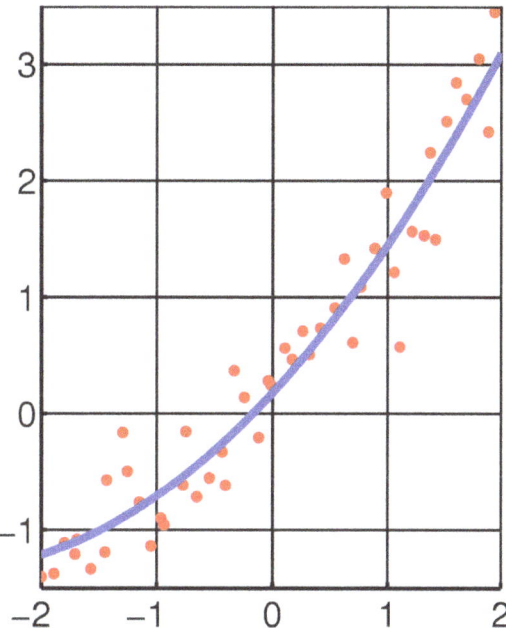

The result of fitting a set of data points with a quadratic function

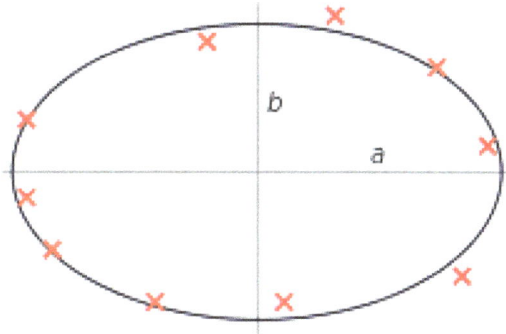

Conic fitting a set of points using least-squares approximation

Least squares problems fall into two categories: linear or ordinary least squares and non-linear least squares, depending on whether or not the residuals are linear in all unknowns. The linear least-squares problem occurs in statistical regression analysis; it has a closed-form solution. The non-linear problem is usually solved by iterative refinement; at each iteration the system is approximated by a linear one, and thus the core calculation is similar in both cases.

Polynomial least squares describes the variance in a prediction of the dependent variable as a function of the independent variable and the deviations from the fitted curve.

When the observations come from an exponential family and mild conditions are satisfied, least-squares estimates and maximum-likelihood estimates are identical. The method of least squares can also be derived as a method of moments estimator.

The following discussion is mostly presented in terms of linear functions but the use of least-squares is valid and practical for more general families of functions. Also, by iteratively applying local quadratic approximation to the likelihood (through the Fisher information), the least-squares method may be used to fit a generalized linear model.

For the topic of approximating a function by a sum of others using an objective function based on squared distances.

The least-squares method is usually credited to Carl Friedrich Gauss (1795), but it was first published by Adrien-Marie Legendre.

History

Context

The method of least squares grew out of the fields of astronomy and geodesy, as scientists and mathematicians sought to provide solutions to the challenges of navigating the Earth's oceans during the Age of Exploration. The accurate description of the behavior of celestial bodies was the key to enabling ships to sail in open seas, where sailors could no longer rely on land sightings for navigation.

The method was the culmination of several advances that took place during the course of the eighteenth century:

- The combination of different observations as being the best estimate of the true value; errors decrease with aggregation rather than increase, perhaps first expressed by Roger Cotes in 1722.

- The combination of different observations taken under the *same* conditions contrary to simply trying one's best to observe and record a single observation accurately. The approach was known as the method of averages. This approach was notably used by Tobias Mayer while studying the librations of the moon in 1750, and by Pierre-Simon Laplace in his work in explaining the differences in motion of Jupiter and Saturn in 1788.

- The combination of different observations taken under *different* conditions. The method came to be known as the method of least absolute deviation. It was notably performed by Roger Joseph Boscovich in his work on the shape of the earth in 1757 and by Pierre-Simon Laplace for the same problem in 1799.

- The development of a criterion that can be evaluated to determine when the solution with the minimum error has been achieved. Laplace tried to specify a mathematical form of the probability density for the errors and define a method of estimation that minimizes the error of estimation. For this purpose, Laplace used a symmetric two-sided exponential distribution we now call Laplace distribution to model the error distribution, and used the sum of absolute deviation as error of estimation. He felt these to be the simplest assumptions he could make, and he had hoped to obtain the arithmetic mean as the best estimate. Instead, his estimator was the posterior median.

The Method

Carl Friedrich Gauss

The first clear and concise exposition of the method of least squares was published by Legendre in 1805. The technique is described as an algebraic procedure for fitting linear equations to data and Legendre demonstrates the new method by analyzing the same data as Laplace for the shape of the earth. The value of Legendre's method of least squares was immediately recognized by leading astronomers and geodesists of the time.

In 1809 Carl Friedrich Gauss published his method of calculating the orbits of celestial bodies. In that work he claimed to have been in possession of the method of least squares since 1795. This naturally led to a priority dispute with Legendre. However, to Gauss's credit, he went beyond Legendre and succeeded in connecting the method of least squares with the principles of probability and to the normal distribution. He had managed to complete Laplace's program of specifying a mathematical form of the probability density for the observations, depending on a finite number of unknown parameters, and define a method of estimation that minimizes the error of estimation. Gauss showed that arithmetic mean is indeed the best estimate of the location parameter by changing both the probability density and the method of estimation. He then turned the problem around by asking what form the density should have and what method of estimation should be used to get the arithmetic mean as estimate of the location parameter. In this attempt, he invented the normal distribution.

An early demonstration of the strength of Gauss' method came when it was used to predict the future location of the newly discovered asteroid Ceres. On 1 January 1801, the Italian astronomer Giuseppe Piazzi discovered Ceres and was able to track its path for 40 days before it was lost in the glare of the sun. Based on these data, astronomers desired to determine the location of Ceres after it emerged from behind the sun without solving Kepler's complicated nonlinear equations of planetary motion. The only predictions that successfully allowed Hungarian astronomer Franz Xaver von Zach to relocate Ceres were those performed by the 24-year-old Gauss using least-squares analysis.

In 1810, after reading Gauss's work, Laplace, after proving the central limit theorem, used it to give a large sample justification for the method of least square and the normal distribution. In 1822,

Gauss was able to state that the least-squares approach to regression analysis is optimal in the sense that in a linear model where the errors have a mean of zero, are uncorrelated, and have equal variances, the best linear unbiased estimator of the coefficients is the least-squares estimator. This result is known as the Gauss–Markov theorem.

The idea of least-squares analysis was also independently formulated by the American Robert Adrain in 1808. In the next two centuries workers in the theory of errors and in statistics found many different ways of implementing least squares.

Problem Statement

The objective consists of adjusting the parameters of a model function to best fit a data set. A simple data set consists of n points (data pairs) (x_i, y_i), i = 1, ..., n, where x_i is an independent variable and y_i is a dependent variable whose value is found by observation. The model function has the form $f(x, \beta)$, where m adjustable parameters are held in the vector β. The goal is to find the parameter values for the model that "best" fits the data. The least squares method finds its optimum when the sum, S, of squared residuals.

$$S = \sum_{i=1}^{n} r_i^2$$

is a minimum. A residual is defined as the difference between the actual value of the dependent variable and the value predicted by the model. Each data point has one residual. Both the sum and the mean of the residuals are equal to zero.

$$r_i = y_i - f(x_i, \beta).$$

An example of a model is that of the straight line in two dimensions. Denoting the y-intercept as β_0 and the slope as β_1, the model function is given by $f(x, \beta) = \beta_0 + \beta_1 x$.

A data point may consist of more than one independent variable. For example, when fitting a plane to a set of height measurements, the plane is a function of two independent variables, x and z, say. In the most general case there may be one or more independent variables and one or more dependent variables at each data point.

Limitations

This regression formulation considers only residuals in the dependent variable. There are two rather different contexts in which different implications apply:

- Regression for prediction. Here a model is fitted to provide a prediction rule for application in a similar situation to which the data used for fitting apply. Here the dependent variables corresponding to such future application would be subject to the same types of observation error as those in the data used for fitting. It is therefore logically consistent to use the least-squares prediction rule for such data.

- Regression for fitting a "true relationship". In standard regression analysis, that leads to fitting by least squares, there is an implicit assumption that errors in the independent variable are zero or strictly controlled so as to be negligible. When errors in the independent variable are non-negligible, models of measurement error can be used; such methods can lead to parameter estimates, hypothesis testing and confidence intervals that take into account the presence of observation errors in the independent variables. An alternative approach is to fit a model by total least squares; this can be viewed as taking a pragmatic approach to balancing the effects of the different sources of error in formulating an objective function for use in model-fitting.

Solving the Least Squares Problem

The minimum of the sum of squares is found by setting the gradient to zero. Since the model contains m parameters, there are m gradient equations:

$$\frac{\partial S}{\partial \beta_j} = 2\sum_i r_i \frac{\partial r_i}{\partial \beta_j} = 0, \, j = 1, \ldots, m,$$

and since $r_i = y_i - f(x_i, \beta)$, the gradient equations become

$$-2\sum_i r_i \frac{\partial f(x_i, \beta)}{\partial \beta_j} = 0, \, j = 1, \ldots, m.$$

The gradient equations apply to all least squares problems. Each particular problem requires particular expressions for the model and its partial derivatives.

Linear Least Squares

A regression model is a linear one when the model comprises a linear combination of the parameters, i.e.,

$$f(x, \beta) = \sum_{j=1}^m \beta_j \phi_j(x),$$

where the function ϕ_j is a function of x.

Letting

$$X_{ij} = \frac{\partial f(x_i, \beta)}{\partial \beta_j} = \phi_j(x_i),$$

we can then see that in that case the least square estimate (or estimator, in the context of a random sample), β is given by

$$\hat{\beta} = (X^T X)^{-1} X^T \mathbf{y}.$$

Non-linear Least Squares

There is, in some cases, a closed-form solution to a non-linear least squares problem – but in general there is not. In the case of no closed-form solution, numerical algorithms are used to find the value of the parameters β that minimizes the objective. Most algorithms involve choosing initial values for the parameters. Then, the parameters are refined iteratively, that is, the values are obtained by successive approximation:

$$\beta_j^{k+1} = \beta_j^k + \Delta\beta_j,$$

where a superscript k is an iteration number, and the vector of increments $\Delta\beta_j$ is called the shift vector. In some commonly used algorithms, at each iteration the model may be linearized by approximation to a first-order Taylor series expansion about β^k:

$$f(x_i, \beta) = f^k(x_i, \beta) + \sum_j \frac{\partial f(x_i, \beta)}{\partial \beta_j}\left(\beta_j - \beta_j^k\right)$$
$$= f^k(x_i, \beta) + \sum_j J_{ij}\Delta\beta_j.$$

The Jacobian J is a function of constants, the independent variable *and* the parameters, so it changes from one iteration to the next. The residuals are given by

$$r_i = y_i - f^k(x_i, \beta) - \sum_{k=1}^m J_{ik}\Delta\beta_k = \Delta y_i - \sum_{j=1}^m J_{ij}\Delta\beta_j.$$

To minimize the sum of squares of r_i, the gradient equation is set to zero and solved for $\Delta\beta_j$:

$$-2\sum_{i=1}^n J_{ij}\left(\Delta y_i - \sum_{k=1}^m J_{ik}\Delta\beta_k\right) = 0,$$

which, on rearrangement, become m simultaneous linear equations, the normal equations:

$$\sum_{i=1}^n \sum_{k=1}^m J_{ij}J_{ik}\Delta\beta_k = \sum_{i=1}^n J_{ij}\Delta y_i \qquad (j = 1, \ldots, m).$$

The normal equations are written in matrix notation as

$$(J^T J)\Delta\beta = J^T \Delta Y.$$

These are the defining equations of the Gauss–Newton algorithm.

Differences between Linear and Nonlinear Least Squares

- The model function, f, in LLSQ (linear least squares) is a linear combination of parameters of the form $f = X_{i1}\beta_1 + X_{i2}\beta_2 + \cdots$ The model may represent a straight line, a parabola or any other linear combination of functions. In NLLSQ (nonlinear least squares) the parameters appear as functions, such as $\beta^2, e^{\beta x}$ and so forth. If the derivatives $\partial f / \partial \beta_j$ are either constant or depend only on the values of the independent variable, the model is linear in the parameters. Otherwise the model is nonlinear.

- Algorithms for finding the solution to a NLLSQ problem require initial values for the parameters, LLSQ does not.

- Like LLSQ, solution algorithms for NLLSQ often require that the Jacobian can be calculated. Analytical expressions for the partial derivatives can be complicated. If analytical expressions are impossible to obtain either the partial derivatives must be calculated by numerical approximation or an estimate must be made of the Jacobian.

- In NLLSQ non-convergence (failure of the algorithm to find a minimum) is a common phenomenon whereas the LLSQ is globally concave so non-convergence is not an issue.

- NLLSQ is usually an iterative process. The iterative process has to be terminated when a convergence criterion is satisfied. LLSQ solutions can be computed using direct methods, although problems with large numbers of parameters are typically solved with iterative methods, such as the Gauss–Seidel method.

- In LLSQ the solution is unique, but in NLLSQ there may be multiple minima in the sum of squares.

- Under the condition that the errors are uncorrelated with the predictor variables, LLSQ yields unbiased estimates, but even under that condition NLLSQ estimates are generally biased.

These differences must be considered whenever the solution to a nonlinear least squares problem is being sought.

Least Squares, Regression Analysis and Statistics

The method of least squares is often used to generate estimators and other statistics in regression analysis.

Consider a simple example drawn from physics. A spring should obey Hooke's law which states that the extension of a spring y is proportional to the force, F, applied to it.

$$y = f(F, k) = kF$$

constitutes the model, where F is the independent variable. To estimate the force constant, k, a series of n measurements with different forces will produce a set of data, $(F_i, y_i), i = 1, \ldots, n,$, where y_i is a measured spring extension. Each experimental observation will contain some error. If we denote this error ε, we may specify an empirical model for our observations,

$$y_i = kF_i + \varepsilon_i.$$

There are many methods we might use to estimate the unknown parameter k. Noting that the n equations in the m variables in our data comprise an overdetermined system with one unknown and n equations, we may choose to estimate k using least squares. The sum of squares to be minimized is

$$S = \sum_{i=1}^{n} (y_i - kF_i)^2.$$

The least squares estimate of the force constant, k, is given by

$$\hat{k} = \frac{\sum_i F_i y_i}{\sum_i F_i^2}.$$

Here it is assumed that application of the force *causes* the spring to expand and, having derived the force constant by least squares fitting, the extension can be predicted from Hooke's law.

In regression analysis the researcher specifies an empirical model. For example, a very common model is the straight line model which is used to test if there is a linear relationship between dependent and independent variable. If a linear relationship is found to exist, the variables are said to be correlated. However, correlation does not prove causation, as both variables may be correlated with other, hidden, variables, or the dependent variable may "reverse" cause the independent variables, or the variables may be otherwise spuriously correlated. For example, suppose there is a correlation between deaths by drowning and the volume of ice cream sales at a particular beach. Yet, both the number of people going swimming and the volume of ice cream sales increase as the weather gets hotter, and presumably the number of deaths by drowning is correlated with the number of people going swimming. Perhaps an increase in swimmers causes both the other variables to increase.

In order to make statistical tests on the results it is necessary to make assumptions about the nature of the experimental errors. A common (but not necessary) assumption is that the errors belong to a normal distribution. The central limit theorem supports the idea that this is a good approximation in many cases.

- The Gauss–Markov theorem. In a linear model in which the errors have expectation zero conditional on the independent variables, are uncorrelated and have equal variances, the best linear unbiased estimator of any linear combination of the observations, is its least-squares estimator. "Best" means that the least squares estimators of the parameters have minimum variance. The assumption of equal variance is valid when the errors all belong to the same distribution.

- In a linear model, if the errors belong to a normal distribution the least squares estimators are also the maximum likelihood estimators.

However, if the errors are not normally distributed, a central limit theorem often nonetheless implies that the parameter estimates will be approximately normally distributed so long as the sample is reasonably large. For this reason, given the important property that the error mean is independent of the independent variables, the distribution of the error term is not an important issue in regression analysis. Specifically, it is not typically important whether the error term follows a normal distribution.

In a least squares calculation with unit weights, or in linear regression, the variance on the jth parameter, denoted $\text{var}(\hat{\beta}_j)$, is usually estimated with

$$\text{var}(\hat{\beta}_j) = \sigma^2 ([X^T X]^{-1})_{jj} \approx \frac{S}{n-m} ([X^T X]^{-1})_{jj},$$

where the true error variance σ^2 is replaced by an estimate based on the minimised value of the sum of squares objective function S. The denominator, $n - m$, is the statistical degrees of freedom.

Confidence limits can be found if the probability distribution of the parameters is known, or an asymptotic approximation is made, or assumed. Likewise statistical tests on the residuals can be made if the probability distribution of the residuals is known or assumed. The probability distribution of any linear combination of the dependent variables can be derived if the probability distribution of experimental errors is known or assumed. Inference is particularly straightforward if the errors are assumed to follow a normal distribution, which implies that the parameter estimates and residuals will also be normally distributed conditional on the values of the independent variables.

Weighted Least Squares

A special case of generalized least squares called weighted least squares occurs when all the off-diagonal entries of Ω (the correlation matrix of the residuals) are null; the variances of the observations (along the covariance matrix diagonal) may still be unequal (heteroscedasticity).

The expressions given above are based on the implicit assumption that the errors are uncorrelated with each other and with the independent variables and have equal variance. The Gauss–Markov theorem shows that, when this is so, $\hat{\beta}$ is a best linear unbiased estimator (BLUE). If, however, the measurements are uncorrelated but have different uncertainties, a modified approach might be adopted. Aitken showed that when a weighted sum of squared residuals is minimized, $\hat{\beta}$ is the BLUE if each weight is equal to the reciprocal of the variance of the measurement.

$$S = \sum_{i=1}^{n} W_{ii} r_i^2, \qquad W_{ii} = \frac{1}{\sigma_i^2}$$

The gradient equations for this sum of squares are

$$-2\sum_{i} W_{ii} \frac{\partial f(x_i, \beta)}{\partial \beta_j} r_i = 0, \qquad j = 1, \ldots, n$$

which, in a linear least squares system give the modified normal equations,

$$\sum_{i=1}^{n} \sum_{k=1}^{m} X_{ij} W_{ii} X_{ik} \hat{\beta}_k = \sum_{i=1}^{n} X_{ij} W_{ii} y_i, \qquad j = 1, \ldots, m.$$

When the observational errors are uncorrelated and the weight matrix, W, is diagonal, these may be written as

$$\left(X^T W X\right)\hat{\beta} = X^T W y.$$

If the errors are correlated, the resulting estimator is the BLUE if the weight matrix is equal to the inverse of the variance-covariance matrix of the observations.

When the errors are uncorrelated, it is convenient to simplify the calculations to factor the weight matrix as $w_{ii} = \sqrt{W_{ii}}$. The normal equations can then be written in the same form as ordinary least squares:

$$\left(X'^{T} X'\right)\hat{\beta} = X'^{T} y'$$

where we define the following scaled matrix and vector:

$$\mathbf{X}' = \mathrm{diag}\left(\mathbf{w}\right)\mathbf{X},$$
$$y' = \mathrm{diag}\left(\mathbf{w}\right)\mathbf{y} = \mathbf{y} \oslash \sigma.$$

This is a type of whitening transformation; the last expression involves an entrywise division.

For non-linear least squares systems a similar argument shows that the normal equations should be modified as follows.

$$(J^{T}WJ)\Delta\beta = J^{T}W\Delta y.$$

Note that for empirical tests, the appropriate W is not known for sure and must be estimated. For this feasible generalized least squares (FGLS) techniques may be used.

Relationship to Principal Components

The first principal component about the mean of a set of points can be represented by that line which most closely approaches the data points (as measured by squared distance of closest approach, i.e. perpendicular to the line). In contrast, linear least squares tries to minimize the distance in the y direction only. Thus, although the two use a similar error metric, linear least squares is a method that treats one dimension of the data preferentially, while PCA treats all dimensions equally.

Regularized Versions

Tikhonov Regularization

In some contexts a regularized version of the least squares solution may be preferable. Tikhonov regularization (or ridge regression) adds a constraint that $\|\beta\|^{2}$, the L_{2}-norm of the parameter vector, is not greater than a given value. Equivalently, it may solve an unconstrained minimization of the least-squares penalty with $\alpha\|\beta\|^{2}$ added, where α is a constant (this is the Lagrangian form of the constrained problem). In a Bayesian context, this is equivalent to placing a zero-mean normally distributed prior on the parameter vector.

Lasso Method

An alternative regularized version of least squares is *Lasso* (least absolute shrinkage and selection operator), which uses the constraint that $\|\beta\|$, the L_{1}-norm of the parameter vector, is no greater than a given value. (As above, this is equivalent to an unconstrained minimization of the least-squares penalty with $\alpha\|\beta\|$ added.) In a Bayesian context, this is equivalent to placing a zero-mean Laplace prior distribution on the parameter vector. The optimization problem may be solved using quadratic programming or more general convex optimization methods, as well as by specific algorithms such as the least angle regression algorithm.

One of the prime differences between Lasso and ridge regression is that in ridge regression, as the penalty is increased, all parameters are reduced while still remaining non-zero, while in Lasso, increasing the penalty will cause more and more of the parameters to be driven to zero. This is an advantage of Lasso over ridge regression, as driving parameters to zero deselects the features from the regression. Thus, Lasso automatically selects more relevant features and discards the others, whereas Ridge regression never fully discards any features. Some feature selection techniques are developed based on the LASSO including Bolasso which bootstraps samples, and FeaLect which analyzes the regression coefficients corresponding to different values of α to score all the features.

The L¹-regularized formulation is useful in some contexts due to its tendency to prefer solutions where more parameters are zero, which gives solutions that depend on fewer variables. For this reason, the Lasso and its variants are fundamental to the field of compressed sensing. An extension of this approach is elastic net regularization.

Linear models play a central part in modern statistical methods. On the one hand, these models are able to approximate a large amount of metric data structures in their entire range of definition or at least piecewise.

Linear Models and Regression Analysis

Suppose the outcome of any process is denoted by a random variable y, called as dependent (or study) variable, depends on k independent (or explanatory) variables denoted by $X_1, X_2, ..., X_k$. Suppose the behaviour of y can be explained by a relationship given by

$$y = f\left(X_1, X_2, ..., X_k, \beta_1, \beta_2, ..., \beta_k\right) + \varepsilon$$

where f is some well defined function and $\beta_1, \beta_2, ..., \beta_k$ are the parameters which characterize the role and contribution of $X_1, X_2, ..., X_k$, respectively. The term ε reflects the stochastic nature of the relationship between y and $X_1, X_2, ..., X_k$ and indicates that such a relationship is not exact in nature. When $\varepsilon = 0$, then the relationship is called the mathematical model otherwise the statistical model. The term "model" is broadly used to represent any phenomenon in a mathematical frame work.

A model or relationship is termed as linear if it is linear in parameters and nonlinear, if it is not linear in parameters. In other words, if all the partial derivatives of y with respect to each of the parameters $\beta_1, \beta_2, ..., \beta_k$ are independent of the parameters, then the model is called as a linear model. If any of the partial derivatives of y with respect to any of the $\beta_1, \beta_2, ..., \beta_k$ is not independent of the parameters, the model is called as nonlinear. Note that the linearity or non-linearity of the model is not described by the linearity or nonlinearity of explanatory variables in the model.

For example

$$y = \beta_1 X_1^2 + \beta_2 \sqrt{X_2} + \beta_3 \log X_3 + \varepsilon$$

is a linear model because $\partial y / \partial \beta_i, (i = 1, 2, 3)$ are independent $\beta_i, (i = 1, 2, 3)$ of the parameters On the other hand,

$$y = \beta_1^2 X_1 + \beta_2 X_2 + \beta_3 \log X + \varepsilon$$

is a nonlinear model because $\partial y / \partial \beta_1 = 2\beta_1 X_1$ depends on β_1 although $\partial y / \partial \beta_2$ and $\partial y / \partial \beta_3$ are independent of any of the β_1, β_2 or β_3.

When the function f is linear in parameters, then $y = f(X_1, X_2, ..., X_k, \beta_1, \beta_2, ..., \beta_k) + \varepsilon$ is called a linear model and when the function f is nonlinear in parameters, then it is called a nonlinear model. In general, the function f is chosen as

$$f(X_1, X_2, ..., X_k, \beta_1, \beta_2, ..., \beta_k) = \beta_1 X_1 + \beta_2 X_2 + ... + \beta_k X_k$$

to describe a linear model. Since $X_1, X_2, ..., X_k$ are pre-determined variables and y is the outcome, so both are known.

Thus the knowledge of the model depends on the knowledge of the parameters $\beta_1, \beta_2, ..., \beta_k$.

The statistical linear modeling essentially consists of developing approaches and tools to determine $\beta_1, \beta_2, ..., \beta_k$ in the linear model

$$y = \beta_1 X_1 + \beta_2 X_2 + ... + \beta_k X_k + \varepsilon$$

given the observations on y and $X_1, X_2, ..., X_k$.

Different statistical estimation procedures, e.g., method of maximum likelihood, principle of least squares, method of moments etc. can be employed to estimate the parameters of the model. The method of maximum likelihood needs further knowledge of the distribution of y whereas the method of moments and the principle of least squares do not need any knowledge about the distribution of y.

The regression analysis is a tool to determine the values of the parameters given the data on y and $X_1, X_2, ..., X_k$. The literal meaning of regression is "to move in the backward direction". Before discussing and understanding the meaning of "backward direction", let us find which of the following statements is correct:

S1: model generates data or

S2: data generates model.

Obviously, S1 is correct. It can be broadly thought that the model exists in nature but is unknown to the experimenter. When some values to the explanatory variables are provided, then the values for the output or study variable are generated accordingly, depending on the form of the function f and the nature of phenomenon. So ideally, the pre-existing model gives rise to the data. Our objective is to determine the functional form of this model. Now we move in the backward direction. We propose to first collect the data on study and explanatory variables. Then we employ some statistical techniques and use this data to know the form of function f. Equivalently, the data from the model is recorded first and then used to determine the parameters of the model. The regression analysis is a technique which helps in determining the statistical model by using the data on study and explanatory variables. The classification of linear and nonlinear regression analysis is based on the determination of linear and nonlinear models, respectively.

Consider a simple example to understand the meaning of "regression". Suppose the yield of crop (y) depends linearly on two explanatory variables, viz., the quantity of a fertilizer (X_1) and level of irrigation (X_2) as

$$y = \beta_1 X_1 + \beta_2 X_2 + \varepsilon.$$

There exist the true values of β_1 and β_2 in nature but are unknown to the experimenter. Some values on y are recorded by providing different values to X_1 and X2. There exists some relationship between y and X_1, X_2 which gives rise to a systematically behaved data on y, X_1 and X_2. Such relationship is unknown to the experimenter. To determine the model, we move in the backward direction in the sense that the collected data is used to determine the unknown parameters β_1 and β_2 of the model. In this sense such an approach is termed as regression analysis.

The theory and fundamentals of linear models lay the foundation for developing the tools for regression analysis that are based on valid statistical theory and concepts.

Steps in Regression Analysis

Regression analysis includes the following steps:

- Statement of the problem under consideration
- Choice of relevant variables
- Collection of data on relevant variables
- Specification of model
- Choice of method for fitting the data
- Fitting of model
- Model validation and criticism
- Using the chosen model(s) for the solution of the posed problem and forecasting.

These steps are examined below.

Statement of the Problem Under Consideration

The first important step in conducting any regression analysis is to specify the problem and the objectives to be addressed by the regression analysis. The wrong formulation or the wrong understanding of the problem will give the wrong statistical inferences. The choice of variables depends upon the objectives of study and understanding of the problem. For example, height and weight of children are related. Now there can be two issues to be addressed.

(i) Determination of height for given weight, or

(ii) determination of weight for given height.

In the case (i), the height is response variable whereas weight is response variable is case (ii). The role of explanatory variables are also interchanged in the cases (i) and (ii).

Choice of Potentially Relevant Variables

Once the problem is carefully formulated and objectives have been decided, the next question is to choose the relevant variables. It has to kept in mind that the correct choice of variables will determine the statistical inferences correctly. For example, in any agricultural experiment, the yield depends on explanatory variables like quantity of fertilizer, rainfall, irrigation, temperature etc. These variables are denoted by $X_1, X_2, ..., X_k$ as a set of k explanatory variables.

Collection of Data on Relevant Variables

Once the objective of study is clearly stated and the variables are chosen, the next question arises is to collect data on such relevant variables. The data is essentially the measurement on these variables. For example, suppose we want to collect the data on age. For this, it is important to know how to record the data on age. Then either the date of birth can be recorded which will provide the exact age on any specific date or the age in terms of completed years as on specific date can be recorded. Moreover, it is also important to decide that whether the data has to be collected on variables as quantitative variables or qualitative variables. For example, if the ages (in years) are 15,17,19,21,23, then these are quantitative values. If the ages are defined by a variable that takes value 1 if ages are less than 18 years and 0 if the ages are more than 18 years, then the earlier recorded data is converted to 1,1,0,0,0. Note that there is a loss of information in converting the quantitative data into qualitative data. The methods and approaches for qualitative and quantitative data are also different. If the study variable is binary, then logistic and probit regressions etc. are used. If all explanatory variables are qualitative, then analysis of variance technique is used. If some explanatory variables are qualitative and others are quantitative, then analysis of covariance technique is used. The techniques of analysis of variance and analysis of covariance are the special cases of regression analysis .

Generally, the data is collected on n subjects, then y denotes the response or study variable and y_1, y_2,..., yn are the n values. If there are k explanatory variables X_1, X_2,..., X_k then x_{ij} denotes the ith value of jth variable, i = 1, 2, ..., n; j = 1, 2,..., k. The observation can be presented in the following table:

Notations for the data used in regression analysis

Observation Number	Response y	Explanatory variables			
		X_1	X_2	$\cdots$	X_k
1	y_1	x_{11}	x_{12}	$\cdots$	x_{1k}
2	y_2	x_{21}	x_{22}	$\cdots$	x_{2k}
3	y_3	x_{31}	x_{32}	$\cdots$	x_{3k}
$\vdots$	$\vdots$	$\vdots$	$\vdots$	$\vdots$	$\vdots$
n	y_n	x_{n1}	x_{n2}	$\cdots$	x_{nk}

Specification of Model

The experimenter or the person working in the subject usually helps in determining the form of the model. Only the form of the tentative model can be ascertained and it will depend on some unknown parameters . For example, a general form will be like

$$y = f(X_1, X_2, ..., X_k; \beta_1, \beta_2, ..., \beta_k) + \varepsilon$$

where ε is the random error reflecting mainly the difference in the observed value of y and the value of y obtained through the model. The form of $f(X_1, X_2, ..., X_k; \beta_1, \beta_2, ..., \beta_k)$ can be linear as well as nonlinear depending on the form of parameters $\beta_1, \beta_2, ..., \beta_k$. A model is said to be linear if it is linear in parameters. For example,

$$y = \beta_1 X_1 + \beta_2 X_1^2 + \beta_3 X_2 + \varepsilon$$
$$y = \beta_1 + \beta_2 \, In \, X_2 + \varepsilon$$

are linear models whereas

$$y = \beta_1 X_1 + \beta_2^2 X_2 + \beta_3 X_2 + \varepsilon$$
$$y = (In \, \beta_1) X_1 + \beta_2 \, X_2 + \varepsilon$$

are non-linear models. Many times, the nonlinear models can be converted into linear models through some transformations. So the class of linear models is wider than what it appears initially.

If a model contains only one explanatory variable, then it is called as simple regression model. When there are more than one independent variables, then it is called as multiple regression model. When there is only one study variable, the regression is termed as univariate regression. When there are more than one study variables, the regression is termed as multivariate regression. Note that the simple and multiple regressions are not same as univariate and multivariate regressions. The simple and multiple regression are determined by the number of explanatory variables whereas univariate and multivariate regressions are determined by the number of study variables.

Choice of Method for Fitting the Data

After the model has been defined and the data have been collected, the next task is to estimate the parameters of the model based on the collected data. This is also referred to as parameter estimation or model fitting. The most commonly used method of estimation is the least squares method. Under certain assumptions, the least squares method produces estimators with desirable properties. The other estimation methods are the maximum likelihood method, ridge method, principal components method etc.

Fitting of Model

The estimation of unknown parameters using appropriate method provides the values of the parameters. Substituting these values in the equation gives us a usable model. This is termed as model fitting. The estimates of parameters $\beta_1, ..., \beta_k$ in the model.

$$y = f(X_1, X_2, ..., X_k, \hat{\beta}_1, \hat{\beta}_2, ..., \hat{\beta}_k) + \varepsilon$$

are denoted as $\hat{\beta}_0, \hat{\beta}_1, ..., \hat{\beta}_k$. which gives the fitted model as

$$y \quad f(X_1, X_2, ..., X, \hat{\beta}_1, \hat{\beta}_2, ..., \hat{\beta})$$

When the value of y is obtained for the given values of X_1, X_2 ,..., X_k, it is denoted as $\hat{y}$ and called as fitted value.

The fitted equation is used for prediction. In this case, $\hat{y}$ is termed as predicted value. Note that the fitted value is where the values used for explanatory variables correspond to one of the n observations in the data whereas predicted value is the one obtained for any set of values of explanatory variables. It is not generally recommended to predict the y - values for the set of those values of explanatory variables which lie outside the range of data. When the values of explanatory variables are the future values of explanatory variables, the predicted values are called forecasted values.

There are different methodologies based on regression analysis. They are described in the following table:

Various Classification of Regression Analysis

Type of Regression	Conditions
Univariate	Only one quantitative response variable
Multivariate	Two or more quantitative response variables
Simple	Only one explanatory variable
Multiple	Two or more explanatory variables
Linear	All parameters enter the equation linearly, possibly after transformation of the data
Nonlinear	The relationship between the response and some of the explanatory variables is nonlinear or some of the parameters appear nonlinearly, but no transformation is possible to make the parameters appear linearly
Analysis of variance	All explanatory variables are qualitative variables
Analysis of Covariance	Some explanatory variables are quantitative variables and

others are qualitative variables

Logistic   The response variable is qualitative

Model Criticism and Selection

The validity of statistical method to be used for regression analysis depends on various assumptions. These assumptions are essentially the assumptions for the model and the data. The quality of statistical inferences heavily depends on whether these assumptions are satisfied or not. For making these assumptions to be valid and to be satisfied, care is needed from beginning of the experiment. One has to be careful in choosing the required assumptions and to examine whether the assumptions are valid for the given experimental conditions or not. It is also important to decide the situations in which the assumptions may not meet.

The validation of the assumptions must be made before drawing any statistical conclusion. Any departure from validity of assumptions will be reflected in the statistical inferences. In fact, the regression analysis is an iterative process where the outputs are used to diagnose, validate, criticize and modify the inputs. The iterative process is illustrated in the following figure.

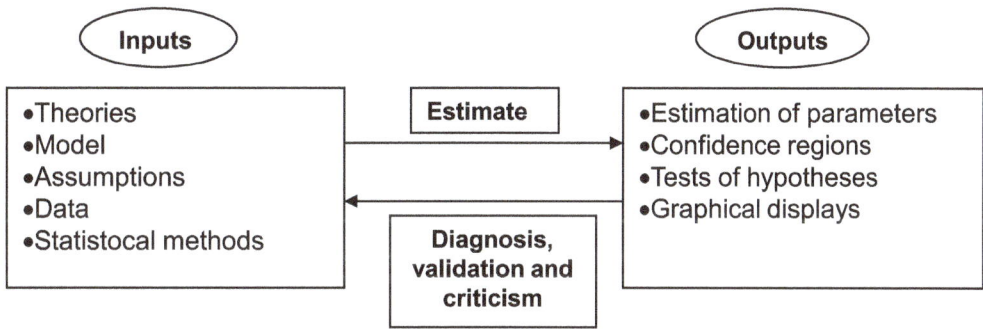

Objectives of Regression Analysis

The determination of explicit form of regression equation is the ultimate objective of regression analysis. It is finally a good and valid relationship between study variable and explanatory variables. The regression equation helps in understanding the interrelationships among the variables. Such regression equation can be used for several purposes. For example, to determine the role of any explanatory variable in the joint relationship in any policy formulation, to forecast the values of response variable for given set of values of explanatory variables.

2

An Overview of Simple Regression Analysis

Regression analysis helps in assessing the relationship between variables. It overlaps with fields like machine learning when used for forecasting or prediction. Variance, covariance, residual sum of squares, ANOVA and prediction interval are some of the topics discussed in the chapter. The topics discussed in the chapter are of great importance to broaden the existing knowledge on regression analysis and linear models.

Simple Linear Regression

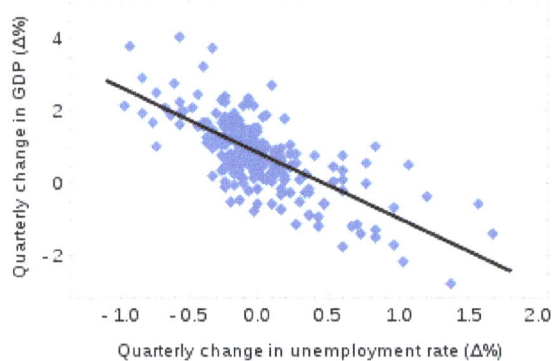

Okun's law in macroeconomics is an example of the simple linear regression. Here the dependent variable (GDP growth) is presumed to be in a linear relationship with the changes in the unemployment rate.

In statistics, simple linear regression is a linear regression model with a single explanatory variable. That is, it concerns two-dimensional sample points with one independent variable and one dependent variable (conventionally, the x and y coordinates in a Cartesian coordinate system) and finds a linear function (a non-vertical straight line) that, as accurately as possible, predicts the dependent variable values as a function of the independent variables. The adjective *simple* refers to the fact that the outcome variable is related to a single predictor.

It is common to make the additional hypothesis that the ordinary least squares method should be used to minimize the *residuals* (vertical distances between the points of the data set and the fitted line). Under this hypothesis, the accuracy of a line through the sample points is measured by the sum of squared residuals, and the goal is to make this sum as small as possible. Other regression methods that can be used in place of ordinary least squares include least absolute deviations (minimizing the sum of absolute values of residuals) and the Theil–Sen estimator (which chooses a line whose slope is the median of the slopes determined by pairs of sample points). Deming regression (total least squares) also finds a line that fits a set of two-dimensional sample points, but (unlike ordinary least squares, least absolute deviations, and median slope regression) it is not really an instance of simple linear regression, because it does not separate the coordinates into one dependent and one independent variable and could potentially return a vertical line as its fit.

The remainder of the article assumes an ordinary least squares regression. In this case, the slope of the fitted line is equal to the correlation between y and x corrected by the ratio of standard deviations of these variables. The intercept of the fitted line is such that it passes through the center of mass $\left(\bar{x}, \bar{y} \right)$ of the data points.

Fitting the Regression Line

Consider the model function

$$y = \alpha + \beta x,$$

which describes a line with slope β and y-intercept α. In an experimental context we may have data points which reflect such a relationship between y and x, but *only approximately*. Say there are n such points and call them $\{(x_i, y_i), i = 1, ..., n\}$. We can describe the approximate relation which exists between y_i and x_i by introducing an error term ε_i to capture the deviation of the data from the model:

$$y_i = \alpha + \beta x_i + \varepsilon_i.$$

This relationship between the parameters and the data points is called a linear regression model, and the ε_i are called "residuals".

The goal is to find values for the parameters α and β which would provide a "best" fit for the data points. As mentioned in the introduction, in this article the "best" fit will be understood as in the least-squares approach: a line that minimizes the sum of squared residuals. In other words, α and β solve the following minimization problem:

$$\text{Find } \min_{\alpha,\beta} Q(\alpha, \beta), \quad \text{for } Q(\alpha, \beta) = \sum_{i=1}^{n} \varepsilon_i^2 = \sum_{i=1}^{n} (y_i - \alpha - \beta x_i)^2$$

By expanding to get a quadratic expression in α and β, we can derive values of α and β that minimize the objective function Q:

$$\alpha = \bar{y} - \hat{\beta}\bar{x},$$

$$\hat{\beta} = \frac{\sum_{i=1}^{n}(x_i - \bar{x})(y_i - \bar{y})}{\sum_{i=1}^{n}(x_i - \bar{x})^2}$$

$$= \frac{\text{Cov}(x, y)}{\text{Var}(x)}$$

$$= r_{xy}\frac{s_y}{s_x},$$

Here we've introduced

- $\bar{x}$ and $\bar{y}$ as the average of the x_i and y_i, respectively

- r_{xy} as the sample correlation coefficient between x and y

- s_x and s_y as the uncorrected sample standard deviations of x and y

- Var and Cov as the (corrected) sample variance and sample covariance, respectively

Substituting the above expressions for $\hat{\alpha}$ and $\hat{\beta}$ into

$$f = \hat{\alpha} + \hat{\beta}x,$$

yields

$$\frac{f - \bar{y}}{s_y} = r_{xy}\frac{x - \bar{x}}{s_x}$$

This shows that r_{xy} is the slope of the regression line of the standardized data points (and that this line passes through the origin).

Generalizing the $\bar{x}$ notation, we can write a horizontal bar over an expression to indicate the average value of that expression over the set of samples. For example:

$$\overline{xy} = \frac{1}{n}\sum_{i=1}^{n} x_i y_i.$$

This notation allows us a concise formula for r_{xy}:

$$r_{xy} = \frac{\overline{xy} - \bar{x}\bar{y}}{\sqrt{\left(\overline{x^2} - \bar{x}^2\right)\left(\overline{y^2} - \bar{y}^2\right)}}$$

The coefficient of determination ("R squared") is equal to r_{xy}^2 when the model is linear with a single independent variable.

Linear Regression Without the Intercept Term

Sometimes it is appropriate to force the regression line to pass through the origin, because x and y are assumed to be proportional. For the model without the intercept term, $y = \beta x$, the OLS estimator for β simplifies to

$$\hat{\beta} = \frac{\sum_{i=1}^{n} x_i y_i}{\sum_{i=1}^{n} x_i^2} = \frac{\overline{xy}}{\overline{x^2}}$$

Substituting $(x - h, y - k)$ in place of (x, y) gives the regression through (h, k):

$$\hat{\beta} = \frac{\overline{(x-h)(y-k)}}{\overline{(x-h)^2}}$$

$$= \frac{\overline{xy} + k\overline{x} - h\overline{y} - hk}{\overline{x^2} - 2h\overline{x} + h^2}$$

$$= \frac{\overline{xy} - \overline{x}\,\overline{y} + (\overline{x} - h)(\overline{y} - k)}{\overline{x^2} - \overline{x}^2 + (\overline{x} - h)^2}$$

$$= \frac{\mathrm{Cov}(x, y) + (\overline{x} - h)(\overline{y} - k)}{\mathrm{Var}(x) + (\overline{x} - h)^2}$$

The last form above demonstrates how moving the line away from the center of mass of the data points affects the slope.

Numerical Properties

1. The regression line goes through the *center of mass* point, $(\overline{x}, \overline{y})$, if the model includes an intercept term (i.e., not forced through the origin).

2. The sum of the residuals is zero if the model includes an intercept term:

$$\sum_{i=1}^{n} \hat{\varepsilon}_i = 0.$$

3. The residuals and x values are uncorrelated, meaning (whether or not there is an intercept term in the model):

$$\sum_{i=1}^{n} x_i \hat{\varepsilon}_i = 0$$

Model-cased Properties

Description of the statistical properties of estimators from the simple linear regression estimates requires the use of a statistical model. The following is based on assuming the validity of a model under which the estimates are optimal. It is also possible to evaluate the properties under other assumptions, such as inhomogeneity, but this is discussed elsewhere.

Unbiasedness

The estimators $\hat{\alpha}$ and $\hat{\beta}$ are unbiased.

To formalize this assertion we must define a framework in which these estimators are random variables. We consider the residuals ε_i as random variables drawn independently from some distribution with mean zero. In other words, for each value of x, the corresponding value of y is generated as a mean response $\alpha + \beta x$ plus an additional random variable ε called the *error term*, equal to zero on average. Under such interpretation, the least-squares estimators $\hat{\alpha}$ and $\hat{\beta}$ will themselves be random variables whose means will equal the "true values" α and β. This is the definition of an unbiased estimator.

Confidence Intervals

The formulas given allow one to calculate the *point estimates* of α and β — that is, the coefficients of the regression line for the given set of data. However, those formulas don't tell us how precise the estimates are, i.e., how much the estimators $\hat{\alpha}$ and $\hat{\beta}$ vary from sample to sample for the specified sample size. Confidence intervals were devised to give a plausible set of values to the estimates one might have if one repeated the experiment a very large number of times.

The standard method of constructing confidence intervals for linear regression coefficients relies on the normality assumption, which is justified if either:

1. the errors in the regression are normally distributed (the so-called *classic regression* assumption), or

2. the number of observations n is sufficiently large, in which case the estimator is approximately normally distributed.

The latter case is justified by the central limit theorem.

Normality Assumption

Under the first assumption above, that of the normality of the error terms, the estimator of the slope coefficient will itself be normally distributed with mean β and variance $\sigma^2 / \sum(x_i - \overline{x})^2$, where σ^2 is the variance of the error terms. At the same time the sum of squared residuals Q is distributed proportionally to χ^2 with $n - 2$ degrees of freedom, and independently from $\hat{\beta}$. This allows us to construct a t-statistic.

$$t = \frac{\hat{\beta} - \beta}{s_{\hat{\beta}}} \sim t_{n-2},$$

where

$$s_{\hat{\beta}} = \sqrt{\frac{\frac{1}{n-2} \sum_{i=1}^{n} \hat{\varepsilon}_i^2}{\sum_{i=1}^{n} (x_i - \overline{x})^2}}$$

is the *standard error* of the estimator $\hat{\beta}$.

This t-statistic has a Student's t-distribution with $n - 2$ degrees of freedom. Using it we can construct a confidence interval for β:

$$\beta \in \left[\hat{\beta} - s_{\hat{\beta}} t^{*}_{n-2}, \hat{\beta} + s_{\hat{\beta}} t^{*}_{n-2} \right],$$

at confidence level $(1 - \gamma)$, where t^{*}_{n-2} is the $\left(1 - \dfrac{\gamma}{2}\right)$-th quantile of the t_{n-2} distribution. For example, if $\gamma = 0.05$ then the confidence level is 95%.

Similarly, the confidence interval for the intercept coefficient α is given by

$$\alpha \in \left[\hat{\alpha} - s_{\hat{\alpha}} t^{*}_{n-2}, \hat{\alpha} + s_{\hat{\alpha}} t^{*}_{n-2} \right],$$

at confidence level $(1 - \gamma)$, where

$$s_{\hat{\alpha}} = s_{\hat{\beta}} \sqrt{\frac{1}{n} \sum_{i=1}^{n} x_i^2} = \sqrt{\frac{1}{n(n-2)} \left(\sum_{j=1}^{n} \hat{\varepsilon}_j^2 \right) \frac{\sum_{i=1}^{n} x_i^2}{\sum_{i=1}^{n} (x_i - \bar{x})^2}}$$

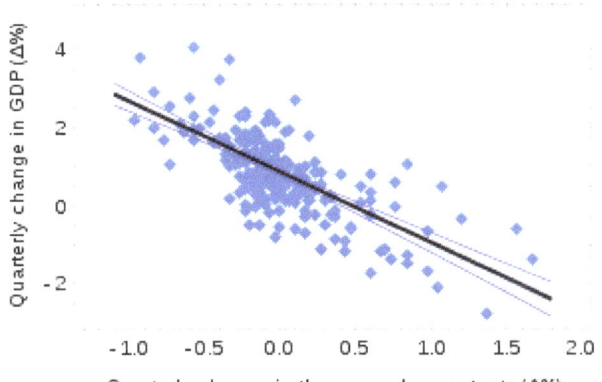

The US "changes in unemployment – GDP growth" regression with the 95% confidence bands.

The confidence intervals for α and β give us the general idea where these regression coefficients are most likely to be. For example, in the *Okun's law* regression the point estimates are

$$\hat{\alpha} = 0.859, \qquad \hat{\beta} = -1.817.$$

The 95% confidence intervals for these estimates are

$$\alpha \in \left[0.76, 0.96\right], \qquad \beta \in \left[-2.06, -1.58\right].$$

In order to represent this information graphically, in the form of the confidence bands around the regression line, one has to proceed carefully and account for the joint distribution of the estimators. It can be shown that at confidence level $(1 - \gamma)$ the confidence band has hyperbolic form given by the equation

$$\hat{y}\big|_{x=\xi} \in \left[\hat{\alpha} + \hat{\beta}\xi \pm t^{*}_{n-2} \sqrt{\left(\frac{1}{n-2} \sum \hat{\varepsilon}_i^2 \right) \cdot \left(\frac{1}{n} + \frac{(\xi - \bar{x})^2}{\sum (x_i - \bar{x})^2} \right)} \right].$$

Asymptotic Assumption

The alternative second assumption states that when the number of points in the dataset is "large enough", the law of large numbers and the central limit theorem become applicable, and then the distribution of the estimators is approximately normal. Under this assumption all formulas derived remain valid, with the only exception that the quantile t^*_{n-2} of Student's t distribution is replaced with the quantile q^* of the standard normal distribution. Occasionally the fraction $\dfrac{1}{n-2}$ is replaced with $\dfrac{1}{n}$. When n is large such a change does not alter the results appreciably.

Numerical Example

This example concerns the data set from the ordinary least squares article. This data set gives average masses for women as a function of their height in a sample of American women of age 30–39. Although the OLS article argues that it would be more appropriate to run a quadratic regression for this data, the simple linear regression model is applied here instead.

Height (m), x_i	1.47	1.50	1.52	1.55	1.57	1.60	1.63	1.65	1.68	1.70	1.73	1.75	1.78	1.80	1.83
Mass (kg), y_i	52.21	53.12	54.48	55.84	57.20	58.57	59.93	61.29	63.11	64.47	66.28	68.10	69.92	72.19	74.46

There are $n = 15$ points in this data set. Hand calculations would be started by finding the following five sums:

$$S_x = \sum x_i = 24.76, \quad S_y = \sum y_i = 931.17$$
$$S_{xx} = \sum x_i^2 = 41.0532, \quad S_{xy} = \sum x_i y_i = 1548.2453, \quad S_{yy} = \sum y_i^2 = 58498.5439$$

These quantities would be used to calculate the estimates of the regression coefficients, and their standard errors.

$$\hat{\beta} = \frac{nS_{xy} - S_x S_y}{nS_{xx} - S_x^2} = 61.272$$

$$\hat{\alpha} = \frac{1}{n}S_y - \hat{\beta}\frac{1}{n}S_x = -39.062$$

$$s_\varepsilon^2 = \frac{1}{n(n-2)}\left[nS_{yy} - S_y^2 - \hat{\beta}^2(nS_{xx} - S_x^2) \right] = 0.5762$$

$$s_{\hat{\beta}}^2 = \frac{ns_\varepsilon^2}{nS_{xx} - S_x^2} = 3.1539$$

$$s_{\hat{\alpha}}^2 = s_{\hat{\beta}}^2 \frac{1}{n}S_{xx} = 8.63185$$

The 0.975 quantile of Student's t-distribution with 13 degrees of freedom is $t^*_{13} = 2.1604$, and thus the 95% confidence intervals for α and β are

$$\alpha \in [\hat{\alpha} \mp t^*_{13} s_\alpha] = [-45.4, -32.7]$$
$$\beta \in [\hat{\beta} \mp t^*_{13} s_\beta] = [57.4, 65.1]$$

The product-moment correlation coefficient might also be calculated:

$$\hat{r} = \frac{nS_{xy} - S_x S_y}{\sqrt{(nS_{xx} - S_x^2)(nS_{yy} - S_y^2)}} = 0.9945$$

This example also demonstrates that sophisticated calculations will not overcome the use of badly prepared data. The heights were originally given in inches, and have been converted to the nearest centimetre. Since the conversion factor is one inch to 2.54 cm, this is *not* a correct conversion. The original inches can be recovered by Round(x/0.0254) and then re-converted to metric: if this is done, the results become

$$\hat{\beta} = 61.6746, \qquad \hat{\alpha} = -39.7468.$$

Thus a seemingly small variation in the data has a real effect.

Derivation of Simple Regression Estimators

We look for $\hat{\alpha}$ and $\hat{\beta}$ that minimize the sum of squared errors (SSE):

$$\min_{\hat{\alpha}, \hat{\beta}} \mathrm{SSE}\left(\hat{\alpha}, \hat{\beta}\right) \equiv \min_{\hat{\alpha}, \hat{\beta}} \sum_{i=1}^{n} \left(y_i - \hat{\alpha} - \hat{\beta} x_i\right)^2$$

To find a minimum take partial derivatives with respect to $\hat{\alpha}$ and $\hat{\beta}$

$$\frac{\partial}{\partial \hat{\alpha}}\left(\mathrm{SSE}\left(\hat{\alpha}, \hat{\beta}\right)\right) = -2\sum_{i=1}^{n}\left(y_i - \hat{\alpha} - \hat{\beta} x_i\right) = 0$$

$$\Rightarrow \sum_{i=1}^{n}\left(y_i - \hat{\alpha} - \hat{\beta} x_i\right) = 0$$

$$\Rightarrow \sum_{i=1}^{n} y_i = \sum_{i=1}^{n} \hat{\alpha} + \hat{\beta}\sum_{i=1}^{n} x_i$$

$$\Rightarrow \sum_{i=1}^{n} y_i = n\hat{\alpha} + \hat{\beta}\sum_{i=1}^{n} x_i$$

$$\Rightarrow \frac{1}{n}\sum_{i=1}^{n} y_i = \hat{\alpha} + \frac{1}{n}\hat{\beta}\sum_{i=1}^{n} x_i$$

$$\Rightarrow \bar{y} = \hat{\alpha} + \hat{\beta}\bar{x}$$

Before taking partial derivative with respect to $\hat{\beta}$, substitute the previous result for $\hat{\alpha}$.

$$\min_{\hat{\alpha},\hat{\beta}} \sum_{i=1}^{n}\left[y_i -\left(\overline{y}-\hat{\beta}\overline{x}\right)-\hat{\beta}x_i\right]^2 = \min_{\hat{\alpha},\hat{\beta}} \sum_{i=1}^{n}\left[\left(y_i-\overline{y}\right)-\hat{\beta}\left(x_i-\overline{x}\right)\right]^2$$

Now, take the derivative with respect to β :

$$\frac{\partial}{\partial\hat{\beta}}\left(SSE\left(\hat{\alpha},\hat{\beta}\right)\right) = -2\sum_{i=1}^{n}\left[\left(y_i-\overline{y}\right)-\hat{\beta}\left(x_i-\overline{x}\right)\right]\left(x_i-\overline{x}\right)=0$$

$$\Rightarrow \sum_{i=1}^{n}\left(y_i-\overline{y}\right)\left(x_i-\overline{x}\right)-\hat{\beta}\sum_{i=1}^{n}\left(x_i-\overline{x}\right)^2 = 0$$

$$\Rightarrow \hat{\beta} = \frac{\sum_{i=1}^{n}\left(y_i-\overline{y}\right)\left(x_i-\overline{x}\right)}{\sum_{i=1}^{n}\left(x_i-\overline{x}\right)^2} = \frac{\text{Cov}(x,y)}{\text{Var}(x)}$$

And finally substitute $\hat{\beta}$ to determine $\hat{\alpha}$

$$\hat{\alpha} = \overline{y}-\hat{\beta}\overline{x}$$

We consider the modeling between the dependent and one independent variable. When there is only one independent variable in the linear regression model, the model is generally termed as simple linear regression model. When there are more than one independent variables in the model, then the linear model is termed as the multiple linear regression model.

Consider a simple linear regression model

$$y = \beta_0 + \beta_1 X + \varepsilon$$

where y is termed as the dependent or study variable and X is termed as independent or explanatory variable. The terms β_0 and β_1 are the parameters of the model. The parameter β_0 is termed as intercept term and the parameter β_1 is termed as slope parameter. These parameters are usually called as regression coefficients. The unobservable error component ε accounts for the failure of data to lie on the straight line and represents the difference between the true and observed realization of y. This is termed as disturbance or error term. There can be several reasons for such difference, e.g., the effect of all deleted variables in the model, variables may be qualitative, inherit randomness in the observations etc. We assume that is observed as independent and identically distributed random variable with mean zero and constant variance σ^2. Later, we will additionally assume that ε is normally distributed.

The independent variable is viewed as controlled by the experimenter, so it is considered as non-stochastic whereas y is viewed as a random variable with

$$E\left(y\right) = \beta_0 + \beta_1 X$$

And

$$Var\left(y\right) = \sigma^2.$$

Sometimes X can also be a random variable. In such a case, instead of simple mean and simple variance of y, we consider the conditional mean of y given X = x as

$$E(y \mid x) = \beta_0 + \beta_1 x$$

and the conditional variance of y given X = x as

$$Var(y \mid x) = \sigma^2$$

When the values of β_0, β_1 and σ^2 are known, the model is completely described.

The parameters β_0, β_1 and σ^2 are generally unknown and ε is unobserved. The determination of the statistical model $y = \beta_0 + \beta_1 X + \varepsilon$ depends on the determination (i.e., estimation) of β_0, β_1 and σ^2.

In order to know the value of the parameters, n pairs of observations $(x_i, y_i)(i=1...,n)$ on (X,y) are observed/collected and are used to determine these unknown parameters.

Various methods of estimation can be used to determine the estimates of the parameters. Among them, the least squares and maximum likelihood principles are the popular methods of estimation.

Suppose a sample of n sets of paired observations $(x_i, y_i)(1 = 1, 2, ..., n)$ are available. These observations are assumed to satisfy the simple linear regression model and so we can write

$$y_i = \beta_0 + \beta_1 x_i + \varepsilon_i (1 = 1, 2, ..., n)$$

The method of least squares estimates the parameters β_0 and β_1 by minimizing the sum of squares of difference between the observations and the line in the scatter diagram. Such an idea is viewed from different perspectives. When the vertical difference between the observations and the line in the scatter diagram is considered and its sum of squares is minimized to obtain the estimates of β_0 and β_1 , the method is known as direct regression.

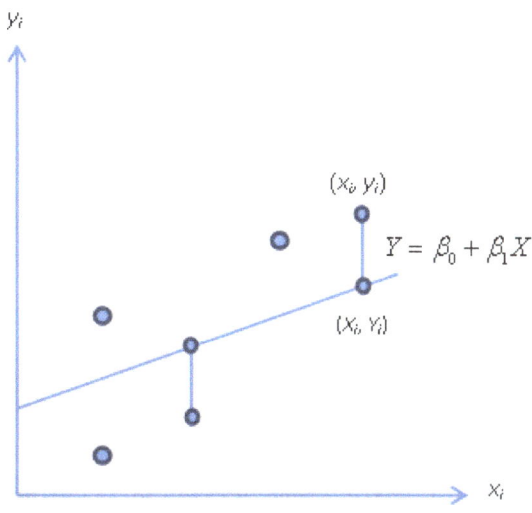

Direct regression method

Alternatively, the sum of squares of difference between the observations and the line in horizontal direction in the scatter diagram can be minimized to obtain the estimates of β_0 and β_1. This is known as reverse (or inverse) regression method.

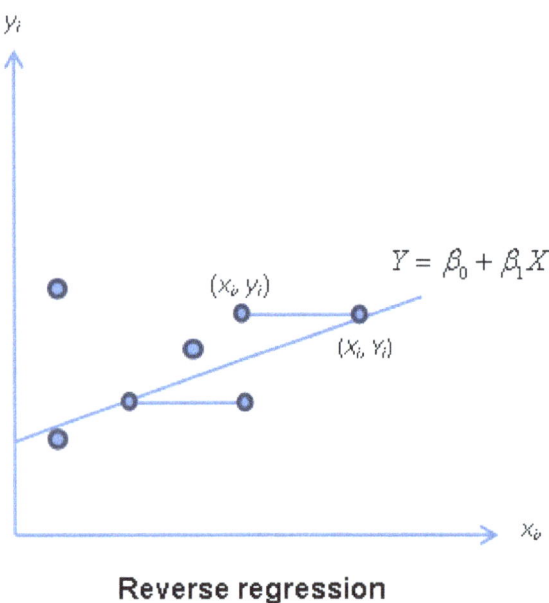

Reverse regression

Instead of horizontal or vertical errors, if the sum of squares of perpendicular distances between the observations and the line in the scatter diagram is minimized to obtain the estimates of β_0 and β_1, the method is known as orthogonal regression or major axis regression method.

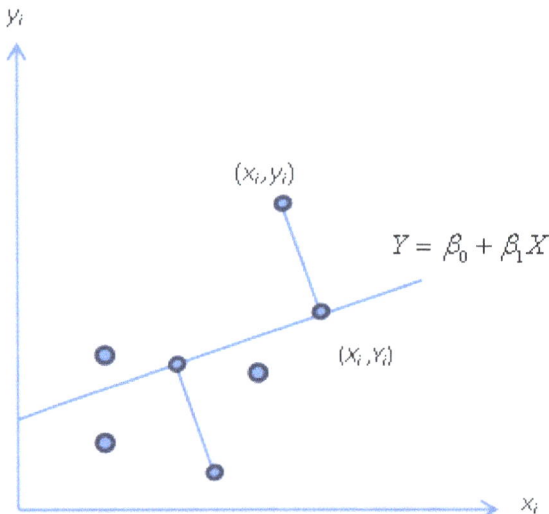

Major axis regression method

Instead of minimizing the distance, the area can also be minimized. The reduced major axis regression method minimizes the sum of the areas of rectangles defined between the observed data points and the nearest point on the line in the scatter diagram to obtain the estimates of regression coefficients. This is shown in the following figure:

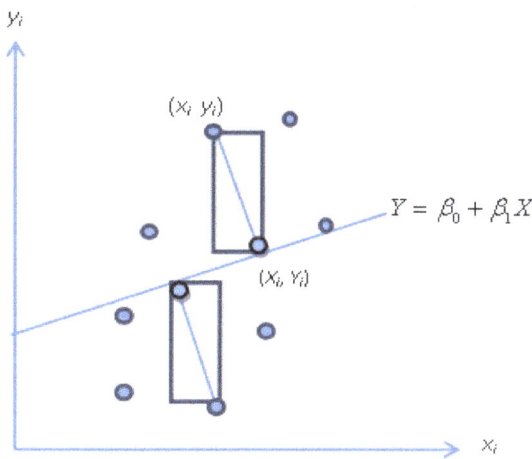

Reduced major axis method

The method of least absolute deviation regression considers the sum of the absolute deviation of the observations from the line in the vertical direction in the scatter diagram as in the case of direct regression to obtain the estimates of β_0 and β_1.

No assumption is required about the form of probability distribution of ε_i in deriving the least squares estimates. For the purpose of deriving the statistical inferences only, we assume that $\varepsilon_i's$ are observed as random variable with

$$E(\varepsilon_i)=0, Var(\varepsilon_i)=\sigma^2 \text{ and } Cov(\varepsilon_i,\varepsilon_j)=0 \text{ of all } i \neq j(i,j=1,2,...,n)$$

This assumption is needed to find the mean, variance and other properties of the least squares estimates. The assumption that $\varepsilon_i's$ are normally distributed is utilized while constructing the tests of hypotheses and confidence intervals of the parameters.

Based on these approaches, different estimates of β_0 and β_1 are obtained which have different statistical properties. Among them the direct regression approach is more popular. Generally, the direct regression estimates are referred as the least squares estimates or ordinary least squares estimates.

Direct Regression Method

This method is also known as the ordinary least squares estimation. Assuming that a set of n paired observations $(x_i, y_i), i=1,2,...,n$ on are available which satisfy the linear regression model $y = \beta_0 + \beta_1 X + \varepsilon$. So we can write the model for each observation as $y_i = \beta_0 + \beta_1 x_i + \varepsilon_i$, $(i=1,2,...,n)$.

The direct regression approach minimizes the sum of squares due to errors given by

$$S(\beta_0,\beta_1) = \sum_{i=1}^{n} \varepsilon_i^2 = \sum_{i=1}^{n}(y_i - \beta_0 - \beta_1 x_i)^2$$

with respect to β_0 and β_1.

The partial derivatives of $S(\beta_0, \beta_1)$ with respect to β_0 are

$$\frac{\partial S(\beta_0, \beta_1)}{\partial \beta_0} = -2 \sum_{i=1}^{n} (y_i - \beta_0 - \beta_1 x_i)$$

and the partial derivative of $S(\beta_0, \beta_1)$ with respect to β_1 is

$$\frac{\partial S(\beta_0, \beta_1)}{\partial \beta_1} = -2 \sum_{i=1}^{n} (y_i - \beta_0 - \beta_1 x_i) x_i$$

The solution of β_0 and β_1 is obtained by setting

$$\frac{\partial S(\beta_0, \beta_1)}{\partial \beta_0} = 0$$

$$\frac{\partial S(\beta_0, \beta_1)}{\partial \beta_1} = 0$$

The solutions of these two equations are called the direct regression estimators, or usually called as the ordinary least squares (OLS) estimators of β_0 and β_1.

This gives the ordinary least squares estimates b_0 of β_0 and b_1 of β_1

$$b_0 = \overline{y} - b_1 \overline{x}$$

$$b_1 = \frac{S_{xy}}{S_{xx}}$$

Where

$$S_{xy} = \sum_{i=1}^{n} (x_i - \overline{x})(y_i - \overline{y}),$$

$$S_{xx} = \sum_{i=1}^{n} (x_i - \overline{x})^2 ,$$

$$\overline{x} = \frac{1}{n} \sum_{i=1}^{n} x_i,$$

$$\overline{y} = \frac{1}{n} \sum_{i=1}^{n} y_i,$$

Further, we have

$$\frac{\partial^2 S(\beta_0,\beta_1)}{\partial \beta_0^2} = -2\sum_{i=1}^{n}(-1) = 2n,$$

$$\frac{\partial^2 S(\beta_0,\beta_1)}{\partial \beta_1^2} = 2\sum_{i=1}^{n}x_i^2$$

$$\frac{\partial^2 S(\beta_0,\beta_1)}{\partial \beta_0 \partial \beta_1} = 2\sum_{i=1}^{n}x_i = 2n\bar{x}.$$

The Hessian matrix which is the matrix of second order partial derivatives in this case is given as

$$\begin{pmatrix} \frac{\partial (\beta_0,\beta_1)}{\partial \beta_0} & \frac{\partial (\beta_0,\beta_1)}{\partial \beta_0 \partial \beta_1} \\ \frac{\partial^2 S(\beta_0,\beta_1)}{\partial \beta_0 \partial \beta_1} & \frac{\partial^2 S(\beta_0,\beta_1)}{\partial} \end{pmatrix}$$

$$\begin{pmatrix} n & n\bar{x} \\ n\bar{x} & \Sigma \end{pmatrix}$$

$$2(\quad)(l,x)$$

where $l = (1,1,...,1)'$ is a n-vector of elements unity $x = (x_1,...,x_n)'$ and is a n-vector of observations on X. The matrix H^* is positive definite if its determinant and the element in the first row and column of H^* are positive.

The determinant of H is given by

$$\left|H^*\right| = 2\left(n\sum_{i=1}^{n}x_i^2 - n^2\bar{x}^2\right)$$

$$= 2n\sum_{i=1}^{n}\left(x_i - \bar{x}\right)^2$$

$$\geq 0$$

The case when $\sum_{i=1}^{n}\left(x_i - \bar{x}\right)^2 = 0$ is not interesting because then all the observations are identical, i.e. $x_i = c$ (some constant). In such a case there is no relationship between x and y in the context of regression analysis. Since $\sum_{i=1}^{n}\left(x_i - \bar{x}\right)^2 > 0$, therefore $\left|H^*\right| > 0$. So H is positive definite for any (β_0,β_1); therefore $S(\beta_0,\beta_1)$ has a global minimum at (b_0,b_1).

The fitted line or the fitted linear regression model is

$$y = b_0 + b_1 x$$

and the predicted values are

$$\hat{y}_i = b_0 + b_1 x_i (i = 1, 2,...,n).$$

The difference between the observed value y_i and the fitted (or predicted) value $\hat{y}_i$ is called as a residual.

The ith residual is

$$e_i = y_i \sim \hat{y}_i \, (i = 1, 2, ..., n).$$

We consider it as

$$
\begin{aligned}
e_i &= y_i - \hat{y}_i \\
&= y_i - \left(b_0 + b_1 x_i\right).
\end{aligned}
$$

Linear Least Squares (Mathematics)

In statistics and mathematics, linear least squares is an approach fitting a mathematical or statistical model to data in cases where the idealized value provided by the model for any data point is expressed linearly in terms of the unknown parameters of the model. The resulting fitted model can be used to summarize the data, to predict unobserved values from the same system, and to understand the mechanisms that may underlie the system.

Mathematically, linear least squares is the problem of approximately solving an overdetermined system of linear equations, where the best approximation is defined as that which minimizes the sum of squared differences between the data values and their corresponding modeled values. The approach is called *linear* least squares since the assumed function is linear in the parameters to be estimated. Linear least squares problems are convex and have a closed-form solution that is unique, provided that the number of data points used for fitting equals or exceeds the number of unknown parameters, except in special degenerate situations. In contrast, non-linear least squares problems generally must be solved by an iterative procedure, and the problems can be non-convex with multiple optima for the objective function. If prior distributions are available, then even an underdetermined system can be solved using the Bayesian MMSE estimator.

In statistics, linear least squares problems correspond to a particularly important type of statistical model called linear regression which arises as a particular form of regression analysis. One basic form of such a model is an ordinary least squares model. The present article concentrates on the mathematical aspects of linear least squares problems, with discussion of the formulation and interpretation of statistical regression models and statistical inferences related to these being dealt with in the articles just mentioned.

Example

As a result of an experiment, four (x, y) data points were obtained, $(1, 6), (2, 5), (3, 7)$ and $(4, 10)$. We hope to find a line $y = \beta_1 + \beta_2 x$ that best fits these four points. In other words, we would like to find the numbers β_1 and β_2 that approximately solve the overdetermined linear system.

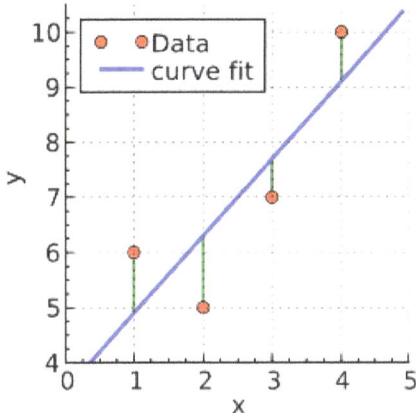

A plot of the data points (in red), the least squares line of best fit (in blue), and the residuals (in green).

$$\beta_1 + 1\beta_2 = 6$$
$$\beta_1 + 2\beta_2 = 5$$
$$\beta_1 + 3\beta_2 = 7$$
$$\beta_1 + 4\beta_2 = 10$$

of four equations in two unknowns in some "best" sense.

The "error", at each point, between the curve fit and the data is the difference between the right- and left-hand sides of the equations above. The least squares approach to solving this problem is to try to make the sum of the squares of these errors as small as possible; that is, to find the minimum of the function

$$S(\beta_1, \beta_2) = [6 - (\beta_1 + 1\beta_2)]^2 + [5 - (\beta_1 + 2\beta_2)]^2$$
$$+ [7 - (\beta_1 + 3\beta_2)]^2 + [10 - (\beta_1 + 4\beta_2)]^2$$
$$= 4\beta_1^2 + 30\beta_2^2 + 20\beta_1\beta_2 - 56\beta_1 - 154\beta_2 + 210.$$

The minimum is determined by calculating the partial derivatives of $S(\beta_1, \beta_2)$ with respect to β_1 and β_2 and setting them to zero

$$\frac{\partial}{\partial} = 0 = 8\beta_1 + 20\beta_2 - 56$$

$$\frac{\partial}{\partial} = 0 = 20\beta_1 + 60\beta_2 - 154.$$

This results in a system of two equations in two unknowns, called the normal equations, which when solved give

$$\beta_1 = 3.5$$
$$\beta_2 = 1.4$$

and the equation $y = 3.5 + 1.4x$ of the line of best fit. The residuals, that is, the discrepancies between the y values from the experiment and the y values calculated using the line of best fit are then found to be $1.1, -1.3, -0.7,$ and 0.9. The minimum value of the sum of squares of the residuals is

$$S(3.5, 1.4) = 1.1^2 + (-1.3)^2 + (-0.7)^2 + 0.9^2 = 4.2.$$

More generally, one can have n regressors x_j, and a linear model

$$y = \beta_1 + \sum_{i=2}^{n+1} \beta_i x_{i-1}.$$

Using a Quadratic Model

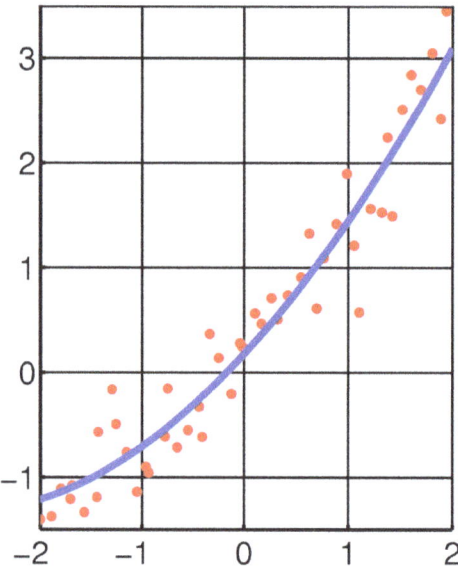

The result of fitting a quadratic function $y = \beta_1 + \beta_2 x + \beta_3 x^2$ (in blue) through a set of data points (x_i, y_i) (in red). In linear least squares the function need not be linear in the argument x, but only in the parameters β_j that are determined to give the best fit.

Importantly, in "linear least squares", we are not restricted to using a line as the model as in the above example. For instance, we could have chosen the restricted quadratic model $y = \beta_1 x^2$. This model is still linear in the β_1 parameter, so we can still perform the same analysis, constructing a system of equations from the data points:

$$6 = \beta_1 (1)^2$$
$$5 = \beta_1 (2)^2$$
$$7 = \beta_1 (3)^2$$
$$10 = \beta_1 (4)^2$$

The partial derivatives with respect to the parameters (this time there is only one) are again computed and set to 0:

$$\frac{\partial S}{\partial \beta_1} = 0 = 708\beta_1 - 498$$

and solved

$$\beta_1 = 0.703$$

leading to the resulting best fit model $y = 0.703x^2$

The General Problem

Consider an overdetermined system

$$\sum_{j=1}^{n} X_{ij}\beta_j = y_i, \ (i = 1, 2, \ldots, m),$$

of m linear equations in n unknown coefficients, $\beta_1, \beta_2, \ldots, \beta_n$, with $m > n$. (Note: for a linear model as above, not all of X contains information on the data points. The first column is populated with ones, $X_{i1} = 1$, only the other columns contain actual data, and n = number of regressors + 1). This can be written in matrix form as

$$X\beta = y,$$

where

$$X = \begin{bmatrix} X_{11} & X_{12} & \cdots & X_{1n} \\ X_{21} & X_{22} & \cdots & X_{2n} \\ \vdots & \vdots & \ddots & \vdots \\ X_{m1} & X_{m2} & \cdots & X_{mn} \end{bmatrix}, \quad \beta = \begin{bmatrix} \beta_1 \\ \beta_2 \\ \vdots \\ \beta_n \end{bmatrix}, \quad y = \begin{bmatrix} y_1 \\ y_2 \\ \vdots \\ y_m \end{bmatrix}.$$

Such a system usually has no solution, so the goal is instead to find the coefficients β which fit the equations "best," in the sense of solving the quadratic minimization problem

$$\hat{\beta} = \underset{\beta}{\arg\min} S(\beta),$$

where the objective function S is given by

$$S(\beta) = \sum_{i=1}^{m} \left| y_i - \sum_{j=1}^{n} X_{ij}\beta_j \right|^2 = \| y - X\beta \|^2 .$$

A justification for choosing this criterion is given in properties below. This minimization problem has a unique solution, provided that the n columns of the matrix X are linearly independent, given by solving the normal equations

$$(X^{T}X)\hat{\beta} = X^{T}y.$$

The matrix $\mathbf{X}^{T}\mathbf{X}$ is known as the Gramian matrix of $\mathbf{X}$, which possesses several nice properties such as being a positive semi-definite matrix, and the matrix $\mathbf{X}^{T}\mathbf{y}$ is known as the moment matrix of regressand by regressors. Finally, $\hat{\beta}$ is the coefficient vector of the least-squares hyperplane, expressed as

$$\hat{\beta} = (\mathbf{X}^{T}\mathbf{X})^{-1}\mathbf{X}^{T}\mathbf{y}.$$

Example Implementation

MATLAB

The following MATLAB code shows implementation of this approach on the data used in the first example above.

```
% MATLAB code for finding the best fit line using least squares method
input = [...              % input in the form of matrix
    1, 6;...              % rows contain points
    2, 5;...
    3, 7;...
    4, 10];
m = length(input);        % number of points
X = [ones(m,1), input(:,1)];   % forming X of X beta = y
y = input(:,2);           % forming y of X beta = y
betaHat = (X' * X) \ (X' * y);   % computing projection of matrix X on
y, giving beta
% display best fit parameters
disp(betaHat);
% plot the best fit line
xx = linspace(0, 5, 2);
yy = betaHat(1) + betaHat(2)*xx;
plot(xx, yy)
% plot the points (data) for which we found the best fit
hold on
plot(input(:,1), input(:,2), 'or')
hold off
```

Python

Python code using the same variable naming as the MATLAB code above:

```
import numpy as np
import matplotlib.pyplot as plt
input = np.array([
    [1, 6],
```

```
    [2, 5],
    [3, 7],
    [4, 10]
])
m = len(input)
X = np.array([np.ones(m), input[:, 0]]).T
y = np.array(input[:, 1]).reshape(-1, 1)
betaHat = np.linalg.solve(X.T.dot(X), X.T.dot(y))
print(betaHat)
plt.figure(1)
xx = np.linspace(0, 5, 2)
yy = np.array(betaHat + betaHat * xx)
plt.plot(xx, yy.T, color='b')
plt.scatter(input[:, 0], input[:, 1], color='r')
plt.show()
```

Julia (Programming Language)

```
using Plots
pyplot() #choose plotting backend
input = [
    1 6
    2 5
    3 7
    4 10]
m = size(input)
X = [ones(m) input[:,1]]
y = input[:,2]
betaHat = X \ y #backslash computes LS-solution (X'X)\X'y (as in Matlab)
print(betaH at)
plot(x->betaHat*x + betaHat,0,5,label="curve fit")
scatter!(input[:,1],input[:,2],label="data")
```

R (Programming Language)

```
m <- 4
n <- 2
input <- matrix(c(1, 6, 2, 5, 3, 7, 4, 10), ncol = n, byrow = T)
k <- rep(1,m)
X <- cbind(k, input[,1])
y <- input[,2]
X.T <- t(X)
betaHat <- solve(X.T%*%X) %*% X.T %*%y
print(betaHat)
plot(input)
abline(betaHat, betaHat)
```

Derivation of the Normal Equations

Define the *ith* residual to be

$$r_i = y_i - \sum_{j=1}^{n} X_{ij}\beta_j.$$

Then S can be rewritten

$$S = \sum_{i=1}^{m} r_i^2.$$

Given that S is convex, it is minimized when its gradient vector is zero (This follows by definition: if the gradient vector is not zero, there is a direction in which we can move to minimize it further.) The elements of the gradient vector are the partial derivatives of S with respect to the parameters:

$$\frac{\partial S}{\partial \beta_j} = 2\sum_{i=1}^{m} r_i \frac{\partial r_i}{\partial \beta_j} \ (j = 1, 2, \ldots, n).$$

The derivatives are

$$\frac{\partial r_i}{\partial \beta_j} = -X_{ij}.$$

Substitution of the expressions for the residuals and the derivatives into the gradient equations gives

$$\frac{\partial S}{\partial \beta_j} = 2\sum_{i=1}^{m} \left(y_i - \sum_{k=1}^{n} X_{ik}\beta_k \right)(-X_{ij}) \ (j = 1, 2, \ldots, n).$$

Thus if $\hat{\beta}$ minimizes S, we have

$$2\sum_{i=1}^{m} \left(y_i - \sum_{k=1}^{n} X_{ik}\hat{\beta}_k \right)(-X_{ij}) = 0 \ (j = 1, 2, \ldots, n).$$

Upon rearrangement, we obtain the normal equations:

$$\sum_{i=1}^{m}\sum_{k=1}^{n} X_{ij}X_{ik}\hat{\beta}_k = \sum_{i=1}^{m} X_{ij}y_i \ (j = 1, 2, \ldots, n).$$

The normal equations are written in matrix notation as

$$(X^TX)\hat{\beta} = X^Ty \ \text{(where } X^T \text{ is the matrix transpose of } X\text{).}$$

The solution of the normal equations yields the vector $\hat{\beta}$ of the optimal parameter values.

Derivation Directly in Terms of Matrices

The normal equations can be derived directly from a matrix representation of the problem as follows. The objective is to minimize

$$S(\beta) = \| y - X\beta \|^2 = (y - X\beta)^\mathsf{T}(y - X\beta) = y^\mathsf{T}y - \beta^\mathsf{T}X^\mathsf{T}y - y^\mathsf{T}X\beta + \beta^\mathsf{T}X^\mathsf{T}X\beta.$$

Note that $: (\beta^\mathsf{T}X^\mathsf{T}y)^\mathsf{T} = y^\mathsf{T}X\beta$ has the dimension 1x1 (the number of columns of $\mathbf{y}$), so it is a scalar and equal to its own transpose, hence $\beta^\mathsf{T}X^\mathsf{T}y = y^\mathsf{T}X\beta$ and the quantity to minimize becomes

$$S(\beta) = y^\mathsf{T}y - 2\beta^\mathsf{T}X^\mathsf{T}y + \beta^\mathsf{T}X^\mathsf{T}X\beta.$$

Differentiating this with respect to β and equating to zero to satisfy the first-order conditions gives

$$-X^\mathsf{T}y + (X^\mathsf{T}X)\beta = 0,$$

which is equivalent to the above-given normal equations. A sufficient condition for satisfaction of the second-order conditions for a minimum is that $\mathbf{X}$ have full column rank, in which case $\mathbf{X}^\mathsf{T}\mathbf{X}$ is positive definite.

Derivation Without Calculus

When $\mathbf{X}^\mathsf{T}\mathbf{X}$ is positive definite, the formula for the minimizing value of β can be derived without the use of derivatives. The quantity

$$S(\beta) = y^T y - 2\beta^T X^T y + \beta^T X^T X\beta$$

can be written as

$$\langle \beta, \beta \rangle - 2\langle \beta, (X^\mathsf{T}X)^{-1}X^\mathsf{T}y \rangle + \langle (X^\mathsf{T}X)^{-1}X^\mathsf{T}y, (X^\mathsf{T}X)^{-1}X^\mathsf{T}y \rangle + C,$$

where C depends only on $\mathbf{y}$ and $\mathbf{X}$, and $\langle .,. \rangle$ is the inner product defined by

$$\langle x, y \rangle = x^\mathsf{T}(X^\mathsf{T}X)y.$$

It follows that $S(\beta)$ is equal to

$$\langle \beta - (X^\mathsf{T}X)^{-1}X^\mathsf{T}y, \beta - (X^\mathsf{T}X)^{-1}X^\mathsf{T}y \rangle + C$$

and therefore minimized exactly when

$$\beta - (X^\mathsf{T}X)^{-1}X^\mathsf{T}y = 0.$$

Generalization for Complex Equations

In general, the coefficients of the matrices X, β and $\mathbf{y}$ can be complex. By using a Hermitian transpose instead of a simple transpose, it is possible to find a vector $\hat{\beta}$ which minimize $S(\beta)$, just as for the real matrices. In order to get the normal equations we follow a similar path as in previous derivations:

$$S(\beta) = \langle y - X\beta, y - X\beta \rangle = \langle y, y \rangle - \overline{\langle X\beta, y \rangle} - \overline{\langle y, X\beta \rangle} + \langle X\beta, X\beta \rangle = y^{\mathrm{T}}\overline{y} - \beta^{\dagger}X^{\dagger}y - y^{\dagger}X\beta + \beta^{\mathrm{T}}X^{\mathrm{T}}\overline{X}\overline{\beta},$$

where $\dagger$ stands for Hermitian transpose.

We should now take derivatives of $S(\beta)$ with respect to each of the coefficient β_j, but first we separate real and imaginary part to deal with the conjugate factors in above expression. For the β_j we have

$$\beta_j = \beta_j^R + i\beta_j^I$$

and the derivatives changes into

$$\frac{\partial S}{\partial \beta_j} = \frac{\partial S}{\partial \beta_j^R}\frac{\partial \beta_j^R}{\partial \beta_j} + \frac{\partial S}{\partial \beta_j^I}\frac{\partial \beta_j^I}{\partial \beta_j} = \frac{\partial S}{\partial \beta_j^R} - i\frac{\partial S}{\partial \beta_j^I} \quad (j = 1, 2, 3, \ldots, n).$$

After rewriting $S(\beta)$ in the summation form and writing β_j explicite, we can calculate both partial derivatives with result:

$$\frac{\partial S}{\partial \beta_j^R} = -\sum_{i=1}^m \left(\overline{X}_{ij}y_i + \overline{y}_i X_{ij} \right) + 2\sum_{i=1}^m X_{ij}\overline{X}_{ij}\beta_j^R + \sum_{i=1}^m \sum_{k \neq j}^n \left(X_{ij}\overline{X}_{ik}\overline{\beta}_k + \beta_k X_{ik}\overline{X}_{ij} \right),$$

$$-i\frac{\partial S}{\partial \beta_j^I} = \sum_{i=1}^m \left(\overline{X}_{ij}y_i - \overline{y}_i X_{ij} \right) - 2i\sum_{i=1}^m X_{ij}\overline{X}_{ij}\beta_j^I + \sum_{i=1}^m \sum_{k \neq j}^n \left(X_{ij}\overline{X}_{ik}\overline{\beta}_k - \beta_k X_{ik}\overline{X}_{ij} \right),$$

which, after adding it together and comparing to zero (minimalization condition for $\hat{\beta}$) yields

$$\sum_{i=1}^m X_{ij}\overline{y}_i = \sum_{i=1}^m \sum_{k=1}^n X_{ij}\overline{X}_{ik}\overline{\hat{\beta}}_k \quad (j = 1, 2, 3, \ldots, n).$$

In matrix form:

$$X^{\mathrm{T}}\overline{y} = X^{\mathrm{T}}\overline{\left(X\hat{\beta} \right)} \quad \text{or} \quad \left(X^{\dagger}X \right)\hat{\beta} = X^{\dagger}y.$$

Computation

A general approach to the least squares problem $\min \| y - X\beta \|^2$ can be described as follows. Suppose that we can find an n by m matrix S such that XS is an orthogonal projection onto the image of X. Then a solution to our minimization problem is given by

$$\beta = Sy$$

simply because

$$X\beta = X(Sy) = (XS)y$$

is exactly a sought for orthogonal projection of **y** onto an image of X. A few popular ways to find such a matrix S are described below.

Inverting the Matrix of the Normal Equations

The algebraic solution of the normal equations can be written as

$$\hat{\beta} = (X^{\mathrm{T}}X)^{-1}X^{\mathrm{T}}y = X^{+}y$$

where X^{+} is the Moore–Penrose pseudoinverse of X. Although this equation is correct and can work in many applications, it is not computationally efficient to invert the normal-equations matrix (the Gramian matrix). An exception occurs in numerical smoothing and differentiation where an analytical expression is required.

If the matrix $X^{\mathrm{T}}X$ is well-conditioned and positive definite, implying that it has full rank, the normal equations can be solved directly by using the Cholesky decomposition $R^{\mathrm{T}}R$, where R is an upper triangular matrix, giving:

$$R^{\mathrm{T}}R\hat{\beta} = X^{\mathrm{T}}y.$$

The solution is obtained in two stages, a forward substitution step, solving for z:

$$R^{\mathrm{T}}\mathbf{z} = X^{\mathrm{T}}\mathbf{y},$$

followed by a backward substitution, solving for $\hat{\beta}$:

$$R\hat{\beta} = z.$$

Both substitutions are facilitated by the triangular nature of R.

Orthogonal Decomposition Methods

Orthogonal decomposition methods of solving the least squares problem are slower than the normal equations method but are more numerically stable because they avoid forming the product $X^{\mathrm{T}}X$.

The residuals are written in matrix notation as

$$r = y - X\hat{\beta}.$$

The matrix X is subjected to an orthogonal decomposition, e.g., the QR decomposition as follows.

$$X = Q\begin{pmatrix} R \\ 0 \end{pmatrix},$$

where Q is an $m \times m$ orthogonal matrix ($Q^{\mathrm{T}}Q=I$) and R is an $n \times n$ upper triangular matrix with $r_{ii} > 0$.

The residual vector is left-multiplied by Q^T.

$$Q^Tr = Q^TY - \left(Q^TQ\right)\begin{pmatrix} R \\ 0 \end{pmatrix}\hat{\beta} = \begin{bmatrix} \left(Q^TY\right)_n - R\hat{\beta} \\ \left(Q^Ty\right)_{m-n} \end{bmatrix} = \begin{bmatrix} u \\ v \end{bmatrix}$$

Because Q is orthogonal, the sum of squares of the residuals, s, may be written as:

$$s = \| \mathbf{r} \|^2 = \mathbf{r}^T\mathbf{r} = \mathbf{r}^TQQ^T\mathbf{r} = \mathbf{u}^T\mathbf{u} + \mathbf{v}^T\mathbf{v}$$

Since $\mathbf{v}$ doesn't depend on β, the minimum value of s is attained when the upper block, $\mathbf{u}$, is zero. Therefore the parameters are found by solving:

$$R\hat{\beta} = \left(Q^Ty\right)_n.$$

These equations are easily solved as R is upper triangular.

An alternative decomposition of X is the singular value decomposition (SVD)

$$X = U\Sigma V^T,$$

where U is m by m orthogonal matrix, V is n by n orthogonal matrix and Σ is an m by n matrix with all its elements outside of the main diagonal equal to 0. The pseudoinverse of Σ is easily obtained by inverting its non-zero diagonal elements and transposing. Hence,

$$\mathbf{XX}^+ = U\Sigma V^TV\Sigma^+U^T = UPU^T,$$

where P is obtained from Σ by replacing its non-zero diagonal elements with ones. Since $\left(\mathbf{XX}^+\right)^* = \mathbf{XX}^+$ (the property of pseudoinverse), the matrix UPU^T is an orthogonal projection onto the image (column-space) of X. In accordance with a general approach described in the introduction above (find XS which is an orthogonal projection),

$$S = \mathbf{X}^+,$$

and thus,

$$\beta = V\Sigma^+U^T\mathbf{y}$$

is a solution of a least squares problem. This method is the most computationally intensive, but is particularly useful if the normal equations matrix, X^TX, is very ill-conditioned (i.e. if its condition number multiplied by the machine's relative round-off error is appreciably large). In that case, including the smallest singular values in the inversion merely adds numerical noise to the solution. This can be cured with the truncated SVD approach, giving a more stable and exact answer, by explicitly setting to zero all singular values below a certain threshold and so ignoring them, a process closely related to factor analysis.

Properties of the Least-squares Estimators

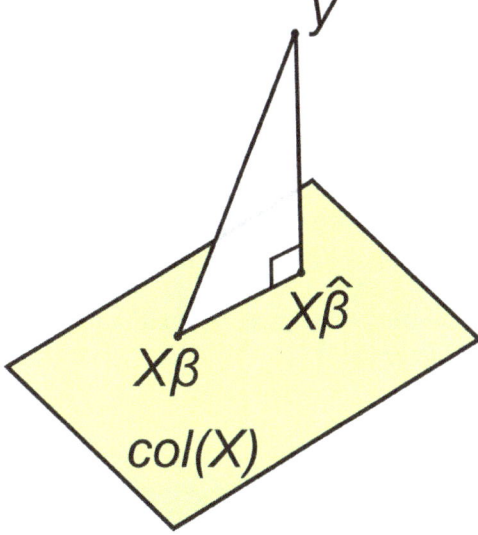

The residual vector, $y - X\hat{\beta}$, which corresponds to the solution of a least squares system, $y = X\hat{\beta} + \epsilon$, is orthogonal to the column space of the matrix X.

The gradient equations at the minimum can be written as

$$(y - X\hat{\beta})^T X = 0.$$

A geometrical interpretation of these equations is that the vector of residuals, $y - X\hat{\beta}$ is orthogonal to the column space of X, since the dot product $(y - X\hat{\beta}) \cdot Xv$ is equal to zero for *any* conformal vector, **v**. This means that $y - X\hat{\beta}$ is the shortest of all possible vectors $y - X\beta$, that is, the variance of the residuals is the minimum possible. This is illustrated at the right.

Introducing $\hat{\gamma}$ and a matrix K with the assumption that a matrix $[X\ K]$ is non-singular and $K^T X$ = 0 (cf. Orthogonal projections), the residual vector should satisfy the following equation:

$$\hat{r} \triangleq y - X\hat{\beta} = K\hat{\gamma}.$$

The equation and solution of linear least squares are thus described as follows:

$$y = \begin{bmatrix} X & K \end{bmatrix} \begin{pmatrix} \hat{\beta} \\ \hat{\gamma} \end{pmatrix},$$

$$\begin{pmatrix} \hat{\beta} \\ \hat{\gamma} \end{pmatrix} = \begin{bmatrix} X & K \end{bmatrix}^{-1} \mathbf{y} = \begin{bmatrix} (X^T X)^{-1} X^T \\ (K^T K)^{-1} K^T \end{bmatrix} y.$$

If the experimental errors, ε, are uncorrelated, have a mean of zero and a constant variance, σ, the Gauss–Markov theorem states that the least-squares estimator, $\hat{\beta}$, has the minimum variance of all estimators that are linear combinations of the observations. In this sense it is the best, or optimal, estimator of the parameters. Note particularly that this property is independent of the statistical distribution function of the errors. In other words, *the distribution function of the errors need not be a normal distribution.* However, for some probability distributions, there is no guarantee that the least-squares solution is even possible given the observations; still, in such cases it is the best estimator that is both linear and unbiased.

For example, it is easy to show that the arithmetic mean of a set of measurements of a quantity is the least-squares estimator of the value of that quantity. If the conditions of the Gauss–Markov theorem apply, the arithmetic mean is optimal, whatever the distribution of errors of the measurements might be.

However, in the case that the experimental errors do belong to a normal distribution, the least-squares estimator is also a maximum likelihood estimator.

These properties underpin the use of the method of least squares for all types of data fitting, even when the assumptions are not strictly valid.

Limitations

An assumption underlying the treatment given above is that the independent variable, x, is free of error. In practice, the errors on the measurements of the independent variable are usually much smaller than the errors on the dependent variable and can therefore be ignored. When this is not the case, total least squares or more generally errors-in-variables models, or *rigorous least squares*, should be used. This can be done by adjusting the weighting scheme to take into account errors on both the dependent and independent variables and then following the standard procedure.

In some cases the (weighted) normal equations matrix X^TX is ill-conditioned. When fitting polynomials the normal equations matrix is a Vandermonde matrix. Vandermonde matrices become increasingly ill-conditioned as the order of the matrix increases. In these cases, the least squares estimate amplifies the measurement noise and may be grossly inaccurate. Various regularization techniques can be applied in such cases, the most common of which is called ridge regression. If further information about the parameters is known, for example, a range of possible values of $\hat{\beta}$, then various techniques can be used to increase the stability of the solution.

Another drawback of the least squares estimator is the fact that the norm of the residuals, $\| y - X\hat{\beta} \|$ is minimized, whereas in some cases one is truly interested in obtaining small error in the parameter $\hat{\beta}$, e.g., a small value of $\| \beta - \hat{\beta} \|$. However, since the true parameter β is necessarily unknown, this quantity cannot be directly minimized. If a prior probability on $\hat{\beta}$ is known, then a Bayes estimator can be used to minimize the mean squared error, $E\{ \| \beta - \hat{\beta} \|^2 \}$. The least squares method is often applied when no prior is known. Surprisingly, when several parameters are being estimated jointly, better estimators can be constructed, an effect known as Stein's phenomenon. For example, if the measurement error is Gaussian, several estimators are known which dominate, or outperform, the least squares technique; the best known of these is the James–Stein estimator. This is an example of more general shrinkage estimators that have been applied to regression problems.

Weighted Linear Least Squares

In some cases the observations may be weighted—for example, they may not be equally reliable. In this case, one can minimize the weighted sum of squares:

$$\underset{\beta}{\arg\min} \sum_{i=1}^{m} w_i \left| y_i - \sum_{j=1}^{n} X_{ij}\beta_j \right|^2 = \underset{\beta}{\arg\min} \left\| W^{1/2}(y - X\beta) \right\|^2.$$

where $w_i > 0$ is the weight of the ith observation, and W is the diagonal matrix of such weights.

The weights should, ideally, be equal to the reciprocal of the variance of the measurement. The normal equations are then:

$$\left(X^\mathrm{T} W X\right)\hat{\beta} = X^\mathrm{T} W \mathbf{y}.$$

This method is used in iteratively reweighted least squares.

Parameter Errors and Correlation

The estimated parameter values are linear combinations of the observed values

$$\hat{\beta} = (X^\mathrm{T} W X)^{-1} X^\mathrm{T} W y.$$

Therefore an expression for the residuals (i.e., the *estimated* errors in the parameters) can be obtained by error propagation from the errors in the observations. Let the variance-covariance matrix for the observations be denoted by M and that of the parameters by M^β. Then

$$M^\beta = (X^\mathrm{T} W X)^{-1} X^\mathrm{T} W M W^\mathrm{T} X (X^\mathrm{T} W^\mathrm{T} X)^{-1}.$$

When $W = M^{-1}$, this simplifies to

$$M^\beta = (X^\mathrm{T} W X)^{-1}.$$

When unit weights are used ($W = I$, the identity matrix), it is implied that the experimental errors are uncorrelated and all equal: $M = \sigma^2 I$, where σ^2 is the *a priori* variance of an observation. In any case, σ^2 is approximated by the reduced chi-squared χ_v^2:

$$M^\beta = \chi_v^2 (X^\mathrm{T} X)^{-1},$$
$$\chi_v^2 = S / v,$$

where S is the minimum value of the (weighted) objective function:

$$S = r^\mathrm{T} W r.$$

The denominator, $v = m - n$, is the number of degrees of freedom.

In all cases, the variance of the parameter β_i is given by M_{ii}^β and the covariance between parameters β_i and β_j is given by M_{ij}^β. Standard deviation is the square root of variance, $\sigma_i = \sqrt{M_{ii}^\beta}$, and the correlation coefficient is given by $\rho_{ij} = M_{ij}^\beta / (\sigma_i \sigma_j)$. These error estimates reflect only random errors in the measurements. The true uncertainty in the parameters is larger due to the presence of systematic errors, which, by definition, cannot be quantified. Note that even though the observations may be uncorrelated, the parameters are typically correlated.

Parameter Confidence Limits

It is often *assumed*, for want of any concrete evidence but often appealing to the central limit theorem—that the error on each observation belongs to a normal distribution with a mean of zero and standard deviation σ. Under that assumption the following probabilities can be derived for a single scalar parameter estimate in terms of its estimated standard error se_β (given here):

68% that the interval $\hat{\beta} \pm se_\beta$ encompasses the true coefficient value

95% that the interval $\hat{\beta} \pm 2se_\beta$ encompasses the true coefficient value

99% that the interval $\hat{\beta} \pm 2.5se_\beta$ encompasses the true coefficient value

The assumption is not unreasonable when $m \gg n$. If the experimental errors are normally distributed the parameters will belong to a Student's t-distribution with $m - n$ degrees of freedom. When $m \gg n$ Student's t-distribution approximates a normal distribution. Note, however, that these confidence limits cannot take systematic error into account. Also, parameter errors should be quoted to one significant figure only, as they are subject to sampling error.

When the number of observations is relatively small, Chebychev's inequality can be used for an upper bound on probabilities, regardless of any assumptions about the distribution of experimental errors: the maximum probabilities that a parameter will be more than 1, 2 or 3 standard deviations away from its expectation value are 100%, 25% and 11% respectively.

Residual Values and Correlation

The residuals are related to the observations by

$$\hat{r} = y - X\hat{\beta} = y - Hy = (I - H)y$$

where H is the idempotent matrix known as the hat matrix:

$$H = X\left(X^{\mathsf{T}}WX\right)^{-1}X^{\mathsf{T}}W$$

and I is the identity matrix. The variance-covariance matrix of the residuals, $M^{\mathbf{r}}$ is given by

$$M^{\mathbf{r}} = (I - H)M(I - H)^{\mathsf{T}}.$$

Thus the residuals are correlated, even if the observations are not.

When $W = M^{-1}$,

$$M^{\mathbf{r}} = (I - H)M.$$

The sum of residual values is equal to zero whenever the model function contains a constant term.

Left-multiply the expression for the residuals by X^T:

$$X^T \hat{r} = X^T \mathbf{y} - X^T X \hat{\beta} = X^T \mathbf{y} - (X^T X)(X^T X)^{-1} X^T \mathbf{y} = 0$$

Say, for example, that the first term of the model is a constant, so that $X_{i1} = 1$ for all i. In that case it follows that

$$\sum_i^m X_{i1}\hat{r}_i = \sum_i^m \hat{r}_i = 0.$$

Thus, in the motivational example, above, the fact that the sum of residual values is equal to zero it is not accidental but is a consequence of the presence of the constant term, a, in the model.

If experimental error follows a normal distribution, then, because of the linear relationship between residuals and observations, so should residuals, but since the observations are only a sample of the population of all possible observations, the residuals should belong to a Student's t-distribution. Studentized residuals are useful in making a statistical test for an outlier when a particular residual appears to be excessively large.

Objective Function

The optimal value of the objective function, found by substituting in the optimal expression for the coefficient vector, can be written as (assuming unweighted observations)

$$S = \mathbf{y}^T (I - H)^T (I - H)\mathbf{y} = \mathbf{y}^T (I - H)\mathbf{y},$$

the latter equality holding, since $(I - H)$ is symmetric and idempotent. It can be shown from this that under an appropriate assignment of weights the expected value of S is $m - n$. If instead unit weights are assumed, the expected value of S is $(m - n)\sigma^2$, where σ^2 is the variance of each observation.

If it is assumed that the residuals belong to a normal distribution, the objective function, being a sum of weighted squared residuals, will belong to a chi-squared (χ^2) distribution with $m - n$ degrees of freedom. Some illustrative percentile values of χ^2 are given in the following table.

$m - n$	$\chi^2_{0.50}$	$\chi^2_{0.95}$	$\chi^2_{0.99}$
10	9.34	18.3	23.2
25	24.3	37.7	44.3
100	99.3	124	136

These values can be used for a statistical criterion as to the goodness of fit. When unit weights are used, the numbers should be divided by the variance of an observation.

Constrained Linear Least Squares

Often it is of interest to solve a linear least squares problem with an additional constraint on the solution. With constrained linear least squares, the original equation

$$X\beta = y$$

must be fit as closely as possible (in the least squares sense) while ensuring that some other property of β is maintained. There are often special-purpose algorithms for solving such problems efficiently. Some examples of constraints are given below:

- Equality constrained least squares: the elements of β must exactly satisfy $L\beta = d$.

- Regularized least squares: the elements of β must satisfy $\|L\beta - y\| \le \alpha$ (choosing α in proportion to the noise standard deviation of **y** prevents over-fitting).

- Non-negative least squares (NNLS): The vector β must satisfy the vector inequality $\beta \ge 0$ defined componentwise—that is, each component must be either positive or zero.

- Box-constrained least squares: The vector β must satisfy the vector inequalities $lb \le \beta \le ub$, each of which is defined componentwise.

- Integer-constrained least squares: all elements of β must be integers (instead of real numbers).

- Phase-constrained least squares: all elements of β must have the same phase (or must be real rather than complex numbers, i.e. phase = 0).

When the constraint only applies to some of the variables, the mixed problem may be solved using separable least squares by letting $\mathbf{X} = [\mathbf{X_1 X_2}]$ and $\beta^T = [\beta_1^T \beta_2^T]$ represent the unconstrained (1) and constrained (2) components. Then substituting the least-squares solution for β_1, i.e.

$$\widehat{\beta_1} = X_1^+(y - X_2\beta_2)$$

back into the original expression gives (following some rearrangement) an equation that can be solved as a purely constrained problem in β_2.

$$PX_2\beta_2 = Py,$$

where $\mathbf{P} := \mathbf{I} - \mathbf{X_1 X_1}^+$ is a projection matrix. Following the constrained estimation of $\widehat{\beta_2}$ the vector $\widehat{\beta_1}$ is obtained from the expression above.

Typical Uses and Applications

- Polynomial fitting: models are polynomials in an independent variable, x:

  ○ Straight line: $f(x, \beta) = \beta_1 + \beta_2 x$.

  ○ Quadratic: $f(x, \beta) = \beta_1 + \beta_2 x + \beta_3 x^2$.

  ○ Cubic, quartic and higher polynomials. For regression with high-order polynomials, the use of orthogonal polynomials is recommended.

- Numerical smoothing and differentiation — this is an application of polynomial fitting.

- Multinomials in more than one independent variable, including surface fitting.

- Curve fitting with B-splines.

- Chemometrics, Calibration curve, Standard addition, Gran plot, analysis of mixtures.

Uses in Data Fitting

The primary application of linear least squares is in data fitting. Given a set of m data points $y_1, y_2, \ldots, y_m$, consisting of experimentally measured values taken at m values $x_1, x_2, \ldots, x_m$ of an independent variable (x_i may be scalar or vector quantities), and given a model function $y = f(x, \beta)$, with $\beta = (\beta_1, \beta_2, \ldots, \beta)$, it is desired to find the parameters β_j such that the model function "best" fits the data. In linear least squares, linearity is meant to be with respect to parameters β_j so

$$f(x, \beta) = \sum_{j=1}^{n} \beta_j \phi_j(x).$$

Here, the functions ϕ_j may be nonlinear with respect to the variable **x**.

Ideally, the model function fits the data exactly, so

$$y_i = f(x_i, \hat{\mathbf{a}})$$

for all $i = 1, 2, \ldots, m$. This is usually not possible in practice, as there are more data points than there are parameters to be determined. The approach chosen then is to find the minimal possible value of the sum of squares of the residuals

$$r_i(\beta) = y_i - f(x_i, \beta), (i = 1, 2, \ldots, m)$$

so to minimize the function

$$S(\beta) = \sum_{i=1}^{m} r_i^2(\beta).$$

After substituting for r_i and then for f, this minimization problem becomes the quadratic minimization problem above with

$$X_{ij} = \phi_j(x_i),$$

and the best fit can be found by solving the normal equations.

Further Discussion

The *numerical methods for linear least squares* are important because linear regression models are among the most important types of model, both as formal statistical models and for exploration of data-sets. The majority of statistical computer packages contain facilities for regression analysis that make use of linear least squares computations. Hence it is appropriate that considerable effort has been devoted to the task of ensuring that these computations are undertaken efficiently and with due regard to round-off error.

Individual statistical analyses are seldom undertaken in isolation, but rather are part of a sequence of investigatory steps. Some of the topics involved in considering numerical methods for linear least squares relate to this point. Thus important topics can be

- Computations where a number of similar, and often nested, models are considered for the same data-set. That is, where models with the same dependent variable but different sets of independent variables are to be considered, for essentially the same set of data-points.

- Computations for analyses that occur in a sequence, as the number of data-points increases.

- Special considerations for very extensive data-sets.

Fitting of linear models by least squares often, but not always, arise in the context of statistical analysis. It can therefore be important that considerations of computation efficiency for such problems extend to all of the auxiliary quantities required for such analyses, and are not restricted to the formal solution of the linear least squares problem.

Rounding Errors

Matrix calculations, like any other, are affected by rounding errors. An early summary of these effects, regarding the choice of computation methods for matrix inversion, was provided by Wilkinson.

Properties of the Direct Regression Estimators

Unbiased Property

Note that $b_1 = \dfrac{S_{xy}}{S_{xx}}$ and $b_0 = \overline{y} - b_1 \overline{x}$ are the linear combinations of $y_i (i = 1, ..., n)$.

Therefore

$$b_1 = \sum_{i=1}^{n} k_i y_i$$

where $k_i = (x_i - \overline{x})/s_{xx}$. Note that $\sum_{i=1}^{n} k_i = 0$ and $\sum_{i=1}^{n} k_i x_i = 1,$

$$E(b_1) = \sum_{i=1}^{n} k_i E(y_i)$$

$$= \sum_{i=1}^{n} k_i (\beta_0 + \beta_1 x_i)$$

$$= \beta_1$$

Thus b_1 is an unbiased estimator of β_1. Next

$$E(b_0) = E\left[\bar{y} - b_1\bar{x}\right]$$
$$= E\left[\beta_0 - \beta_1\bar{x} - b_1\bar{x}\right]$$
$$= \beta_0 - \beta_1\bar{x} - b_1\bar{x}$$
$$= \beta_0$$

Thus b_0 is an unbiased estimators of β_0.

Variance

In probability theory and statistics, variance is the expectation of the squared deviation of a random variable from its mean, and it informally measures how far a set of (random) numbers are spread out from their mean. The variance has a central role in statistics. It is used in descriptive statistics, statistical inference, hypothesis testing, goodness of fit, and Monte Carlo sampling, amongst many others. This makes it a central quantity in numerous fields such as physics, biology, chemistry, cryptography, economics, and finance. The variance is the square of the standard deviation, the second central moment of a distribution, and the covariance of the random variable with itself, and it is often represented by σ^2, s^2, or $Var(X)$.

Definition

The variance of a random variable X is the expected value of the squared deviation from the mean of $X, \mu = E[X]$:

$$\mathrm{Var}(X) = \mathrm{E}\left[(X - \mu)^2\right].$$

This definition encompasses random variables that are generated by processes that are discrete, continuous, neither, or mixed. The variance can also be thought of as the covariance of a random variable with itself:

$$\mathrm{Var}(X) = \mathrm{Cov}(X, X).$$

The variance is also equivalent to the second cumulant of a probability distribution that generates X. The variance is typically designated as $\mathrm{Var}(X)$, σ_X^2, or simply σ^2 (pronounced "sigma squared"). The expression for the variance can be expanded:

$$\mathrm{Var}(X) = \mathrm{E}\left[(X - \mathrm{E}[X])^2\right]$$
$$= \mathrm{E}\left[X^2 - 2X\,\mathrm{E}[X] + \mathrm{E}[X]^2\right]$$
$$= \mathrm{E}\left[X^2\right] - 2\,\mathrm{E}[X]\mathrm{E}[X] + \mathrm{E}[X]^2$$
$$= \mathrm{E}\left[X^2\right] - \mathrm{E}[X]^2$$

A mnemonic for the above expression is "mean of square minus square of mean". On computational floating point arithmetic, this equation should not be used, because it suffers from catastrophic cancellation if the two components of the equation are similar in magnitude. There exist numerically stable alternatives.

Continuous Random Variable

If the random variable X represents samples generated by a continuous distribution with probability density function $f(x)$, then the population variance is given by

$$\text{Var}(X) = \sigma^2 = \int (x-\mu)^2 f(x)dx = \int x^2 f(x)dx - 2\mu \int x f(x)dx + \int \mu^2 f(x)dx = \int x^2 f(x)dx - \mu^2$$

where μ is the expected value

$$\mu = \int x f(x)dx$$

and where the integrals are definite integrals taken for x ranging over the range of X.

If a continuous distribution does not have an expected value, as is the case for the Cauchy distribution, it does not have a variance either. Many other distributions for which the expected value does exist also do not have a finite variance because the integral in the variance definition diverges. An example is a Pareto distribution whose index k satisfies $1 < k \le 2$.

Discrete Random Variable

If the generator of random variable X is discrete with probability mass function $x_1 \mapsto p_1, x_2 \mapsto p_2, \ldots, x_n \mapsto p_n$ then

$$\text{Var}(X) = \sum_{i=1}^{n} p_i \cdot (x_i - \mu)^2,$$

or equivalently

$$\text{Var}(X) = \sum_{i=1}^{n} p_i x_i^2 - \mu^2,$$

where μ is the average value, i.e.

$$\mu = \sum_{i=1}^{n} p_i \cdot x_i.$$

(When such a discrete weighted variance is specified by weights whose sum is not 1, then one divides by the sum of the weights.)

The variance of a set of n equally likely values can be written as

$$\text{Var}(X) = \frac{1}{n} \sum_{i=1}^{n} (x_i - \mu)^2,$$

where μ is the expected value, i.e.,

$$\mu = \frac{1}{n} \sum_{i=1}^{n} x_i.$$

The variance of a set of n equally likely values can be equivalently expressed, without directly referring to the mean, in terms of squared deviations of all points from each other:

$$\mathrm{Var}(X) = \frac{1}{n^2} \sum_{i=1}^{n} \sum_{j=1}^{n} \frac{1}{2} (x_i - x_j)^2 = \frac{1}{n^2} \sum_{i} \sum_{j>i} (x_i - x_j)^2.$$

Examples

Normal Distribution

The normal distribution with parameters μ and σ is a continuous distribution whose probability density function is given by

$$f(x) = \frac{1}{\sqrt{2\pi\sigma^2}} e^{-\frac{(x-\mu)^2}{2\sigma^2}}.$$

In this distribution, $\mathrm{E}[X] = \mu$ and the variance $\mathrm{Var}(X)$ is related with σ via

$$\mathrm{Var}(X) = \int_{-\infty}^{\infty} \frac{x^2}{\sqrt{2\pi\sigma^2}} e^{-\frac{(x-\mu)^2}{2\sigma^2}} dx - \mu^2 = \sigma^2.$$

The role of the normal distribution in the central limit theorem is in part responsible for the prevalence of the variance in probability and statistics.

Exponential Distribution

The exponential distribution with parameter λ is a continuous distribution whose support is the semi-infinite interval $[0, \infty)$. Its probability density function is given by

$$f(x) = \lambda e^{-\lambda x}$$

and it has expected value $\mu = \lambda^{-1}$. The variance is equal to

$$\mathrm{Var}(X) = \int_{0}^{\infty} x^2 \lambda e^{-\lambda x} dx - \mu^2 = \lambda^{-2}.$$

So for an exponentially distributed random variable, $\sigma^2 = \mu^2$.

Poisson Distribution

The Poisson distribution with parameter λ is a discrete distribution for $k = 0, 1, 2, \ldots$ Its probability mass function is given by

$$p(k) = \frac{\lambda^k}{k!} e^{-\lambda},$$

and it has expected value $\mu = \lambda$. The variance is equal to

$$\text{Var}(X) = \left(\sum_{k=0}^{\infty} k^2 \frac{\lambda^k}{k!} e^{-\lambda} \right) - \mu^2 = \lambda,$$

So for a Poisson-distributed random variable, $\sigma^2 = \mu..$

Binomial Distribution

The binomial distribution with parameters n and p is a discrete distribution for $k = 0, 1, 2, \ldots, n$. Its probability mass function is given by

$$p(k) = \binom{n}{k} p^k (1-p)^{n-k},$$

and it has expected value $\mu = np$. The variance is equal to

$$\text{Var}(X) = \left(\sum_{k=0}^{n} k^2 \binom{n}{k} p^k (1-p)^{n-k} \right) - \mu^2 = np(1-p).$$

As a simple example, the binomial distribution with $p = 1/2$ describes the probability of getting k heads in n tosses. Thus the expected value of the number of heads is $\frac{n}{2}$, and the variance is $\frac{n}{4}$.

Fair Die

A six-sided fair die can be modelled with a discrete random variable with outcomes 1 through 6, each with equal probability $\frac{1}{6}$. The expected value is $\frac{1+2+3+4+5+6}{6} = 3.5$. Therefore, the variance can be computed to be

$$\sum_{i=1}^{6} \tfrac{1}{6}(i-3.5)^2 = \tfrac{1}{6} \sum_{i=1}^{6} (i-3.5)^2 = \tfrac{1}{6} \left((-2.5)^2 + (-1.5)^2 + (-0.5)^2 + 0.5^2 + 1.5^2 + 2.5^2 \right)$$

$$= \tfrac{1}{6} \cdot 17.50 = \tfrac{35}{12} \approx 2.92.$$

The general formula for the variance of the outcome X of a die of n sides is

$$\sigma^2 = E(X^2) - (E(X))^2 = \frac{1}{n} \sum_{i=1}^{n} i^2 - \left(\frac{1}{n} \sum_{i=1}^{n} i \right)^2$$

$$= \tfrac{1}{6}(n+1)(2n+1) - \tfrac{1}{4}(n+1)^2$$

$$= \frac{n^2 - 1}{12}.$$

Properties

Basic Properties

Variance is non-negative because the squares are positive or zero:

$$\mathrm{Var}(X) \geq 0.$$

The variance of a constant random variable is zero, and if the variance of a variable in a data set is 0, then all the entries have the same value:

$$P(X = a) = 1 \Leftrightarrow \mathrm{Var}(X) = 0.$$

Variance is invariant with respect to changes in a location parameter. That is, if a constant is added to all values of the variable, the variance is unchanged:

$$\mathrm{Var}(X + a) = \mathrm{Var}(X).$$

If all values are scaled by a constant, the variance is scaled by the square of that constant:

$$\mathrm{Var}(aX) = a^2 \, \mathrm{Var}(X).$$

The variance of a sum of two random variables is given by

$$\mathrm{Var}(aX + bY) = a^2 \, \mathrm{Var}(X) + b^2 \, \mathrm{Var}(Y) + 2ab\mathrm{Cov}(X, Y),$$
$$\mathrm{Var}(aX - bY) = a^2 \, \mathrm{Var}(X) + b^2 \, \mathrm{Var}(Y) - 2ab\mathrm{Cov}(X, Y),$$

where $\mathrm{Cov}(\cdot, \cdot)$ is the covariance. In general we have for the sum of N random variables $\{X_1, \ldots, X_N\}$:

$$\mathrm{Var}\left(\sum_{i=1}^{N} X_i\right) = \sum_{i,j=1}^{N} \mathrm{Cov}(X_i, X_j) = \sum_{i=1}^{N} \mathrm{Var}(X_i) + \sum_{i \neq j} \mathrm{Cov}(X_i, X_j).$$

These results lead to the variance of a linear combination as:

$$\mathrm{Var}\left(\sum_{i=1}^{N} a_i X_i\right) = \sum_{i,j=1}^{N} a_i a_j \, \mathrm{Cov}(X_i, X_j)$$
$$= \sum_{i=1}^{N} a_i^2 \, \mathrm{Var}(X_i) + \sum_{i \neq j} a_i a_j \, \mathrm{Cov}(X_i, X_j)$$
$$= \sum_{i=1}^{N} a_i^2 \, \mathrm{Var}(X_i) + 2 \sum_{1 \leq i < j \leq N} a_i a_j \, \mathrm{Cov}(X_i, X_j).$$

If the random variables $X_1, \ldots, X_N$ are such that

$$\mathrm{Cov}(X_i, X_j) = 0 \,, \forall \, (i \neq j),$$

they are said to be uncorrelated. It follows immediately from the expression given earlier that if the random variables $X_1,\ldots,X_N$ are uncorrelated, then the variance of their sum is equal to the sum of their variances, or, expressed symbolically:

$$\mathrm{Var}\left(\sum_{i=1}^{N} X_i\right) = \sum_{i=1}^{N} \mathrm{Var}(X_i).$$

Since independent random variables are always uncorrelated, the equation above holds in particular when the random variables $X_1,\ldots,X_n$ are independent. Thus independence is sufficient but not necessary for the variance of the sum to equal the sum of the variances.

Sum of Uncorrelated Variables (Bienaymé Formula)

One reason for the use of the variance in preference to other measures of dispersion is that the variance of the sum (or the difference) of uncorrelated random variables is the sum of their variances:

$$\mathrm{Var}\left(\sum_{i=1}^{n} X_i\right) = \sum_{i=1}^{n} \mathrm{Var}(X_i).$$

This statement is called the Bienaymé formula and was discovered in 1853. It is often made with the stronger condition that the variables are independent, but being uncorrelated suffices. So if all the variables have the same variance σ^2, then, since division by n is a linear transformation, this formula immediately implies that the variance of their mean is

$$\mathrm{Var}\left(\bar{X}\right) = \mathrm{Var}\left(\frac{1}{n}\sum_{i=1}^{n} X_i\right) = \frac{1}{n^2}\sum_{i=1}^{n} \mathrm{Var}\left(X_i\right) = \frac{\sigma^2}{n}.$$

That is, the variance of the mean decreases when n increases. This formula for the variance of the mean is used in the definition of the standard error of the sample mean, which is used in the central limit theorem.

To prove the initial statement, it suffices to show that

$$\mathrm{Var}(X+Y) = \mathrm{Var}(X) + \mathrm{Var}(Y).$$

The general result then follows by induction. Starting with the definition,

$$\begin{aligned}
\mathrm{Var}(X+Y) &= E[(X+Y)^2] - (E[X+Y])^2 \\
&= E[X^2 + 2XY + Y^2] - (E[X] + E[Y])^2.
\end{aligned}$$

Using the linearity of the expectation operator and the assumption of independence (or uncorrelatedness) of X and Y, this further simplifies as follows:

$$\begin{aligned}
\mathrm{Var}(X+Y) &= E[X^2] + 2E[XY] + E[Y^2] - (E[X]^2 + 2E[X]E[Y] + E[Y]^2) \\
&= E[X^2] + E[Y^2] - E[X]^2 - E[Y]^2 \\
&= \mathrm{Var}(X) + \mathrm{Var}(Y).
\end{aligned}$$

Sum of Correlated Variables

In general, if the variables are correlated, then the variance of their sum is the sum of their covariances:

$$\text{Var}\left(\sum_{i=1}^{n} X_i\right) = \sum_{i=1}^{n}\sum_{j=1}^{n} \text{Cov}(X_i, X_j) = \sum_{i=1}^{n} \text{Var}(X_i) + 2\sum_{1 \le i < j \le n} \text{Cov}(X_i, X_j).$$

(Note: The second equality comes from the fact that $\text{Cov}(X_i, X_i) = \text{Var}(X_i)$.)

Here $\text{Cov}(\cdot, \cdot)$ is the covariance, which is zero for independent random variables (if it exists). The formula states that the variance of a sum is equal to the sum of all elements in the covariance matrix of the components. The next expression states equivalently that the variance of the sum is the sum of the diagonal of covariance matrix plus two times the sum of its upper triangular elements (or its lower triangular elements); this emphasizes that the covariance matrix is symmetric. This formula is used in the theory of Cronbach's alpha in classical test theory.

So if the variables have equal variance σ^2 and the average correlation of distinct variables is ρ, then the variance of their mean is

$$\text{Var}(\bar{X}) = \frac{\sigma^2}{n} + \frac{n-1}{n}\rho\sigma^2.$$

This implies that the variance of the mean increases with the average of the correlations. In other words, additional correlated observations are not as effective as additional independent observations at reducing the uncertainty of the mean. Moreover, if the variables have unit variance, for example if they are standardized, then this simplifies to

$$\text{Var}(\bar{X}) = \frac{1}{n} + \frac{n-1}{n}\rho.$$

This formula is used in the Spearman–Brown prediction formula of classical test theory. This converges to ρ if n goes to infinity, provided that the average correlation remains constant or converges too. So for the variance of the mean of standardized variables with equal correlations or converging average correlation we have

$$\lim_{n \to \infty} \text{Var}(\bar{X}) = \rho.$$

Therefore, the variance of the mean of a large number of standardized variables is approximately equal to their average correlation. This makes clear that the sample mean of correlated variables does not generally converge to the population mean, even though the Law of large numbers states that the sample mean will converge for independent variables.

Matrix Notation for the Variance of A Linear Combination

Define X as a column vector of n random variables $X_1, \ldots, X_n$, and c as a column vector of n scalars $c_1, \ldots, c_n$. Therefore, $c^T X$ is a linear combination of these random variables, where c^T denotes the transpose of c. Also let Σ be the covariance matrix of X. The variance of $c^T X$ is then given by:

$$\text{Var}(c^T X) = c^T \Sigma c.$$

Weighted Sum of Variables

The scaling property and the Bienaymé formula, along with the property of the covariance $\text{Cov}(aX, bY) = ab\,\text{Cov}(X, Y)$ jointly imply that

$$\text{Var}(aX \pm bY) = a^2\,\text{Var}(X) + b^2\,\text{Var}(Y) \pm 2ab\text{Cov}(X,Y).$$

This implies that in a weighted sum of variables, the variable with the largest weight will have a disproportionally large weight in the variance of the total. For example, if X and Y are uncorrelated and the weight of X is two times the weight of Y, then the weight of the variance of X will be four times the weight of the variance of Y.

The expression above can be extended to a weighted sum of multiple variables:

$$\text{Var}\left(\sum_i^n a_i X_i\right) = \sum_{i=1}^n a_i^2\,\text{Var}(X_i) + 2\sum\sum_{1\le i\,<j\le n} a_i a_j\,\text{Cov}(X_i, X_j)$$

Product of Independent Variables

If two variables X and Y are independent, the variance of their product is given by

$$\text{Var}(XY) = [E(X)]^2\,\text{Var}(Y) + [E(Y)]^2\,\text{Var}(X) + \text{Var}(X)\,\text{Var}(Y).$$

Equivalently, using the basic properties of expectation, it is given by

$$\text{Var}(XY) = E(X^2)E(Y^2) - [E(X)]^2[E(Y)]^2.$$

Product of Statistically Dependent Variables

In general, if two variables are statistically dependent, the variance of their product is given by:

$$\begin{aligned}
\text{Var}(XY) &= E[X^2 Y^2] - [E(XY)]^2 \\
&= \text{Cov}(X^2, Y^2) + E(X^2)E(Y^2) - [E(XY)]^2 \\
&= \text{Cov}(X^2, Y^2) + (\text{Var}(X) + [E(X)]^2)(\text{Var}(Y) + [E(Y)]^2) - [\text{Cov}(X,Y) + E(X)E(Y)]^2
\end{aligned}$$

Decomposition

The general formula for variance decomposition or the law of total variance is: If X and Y are two random variables, and the variance of X exists, then

$$\text{Var}[X] = E_Y(\text{Var}[X \mid Y]) + \text{Var}_Y(E[X \mid Y]).$$

where $E(X \mid Y)$ is the conditional expectation of X given Y, and $Var(X \mid Y)$ is the conditional variance of X given Y. (A more intuitive explanation is that given a particular value of y, then X follows a distribution with mean $E(X \mid Y)$ and variance $Var(X \mid Y)$). As $E(X \mid Y)$ is a function of the variable Y, the outer expectation or variance is taken with respect to Y. The above formula tells how to find $Var(X)$ based on the distributions of these two quantities when Y is allowed to vary.

In particular, if Y is a discrete random variable assuming $y_1, y_2, \ldots, y_n$ with corresponding probability masses $p_1, p_2, \ldots, p_n$, then in the formula for total variance, the first term on the right-hand side becomes

$$E_Y(Var[X \mid Y]) = \sum_{i=1}^{n} p_i \sigma_i^2,$$

where $\sigma_i^2 = Var[X \mid y_i]$. Similarly, the second term on the right-hand side becomes

$$Var_Y(E[X \mid Y]) = \sum_{i=1}^{n} p_i \mu_i^2 - \left(\sum_{i=1}^{n} p_i \mu_i\right)^2 = \sum_{i=1}^{n} p_i \mu_i^2 - \mu^2,$$

where $\mu_i = E[X \mid y_i]$ and $\mu = \sum_{i=1}^{n} p_i \mu_i$. Thus the total variance is given by

$$Var[X] = \sum_{i=1}^{n} p_i \sigma_i^2 + \left(\sum_{i=1}^{n} p_i \mu_i^2 - \mu^2\right).$$

A similar formula is applied in analysis of variance, where the corresponding formula is

$$MS_{total} = MS_{between} + MS_{within};$$

here MS refers to the Mean of the Squares. In linear regression analysis the corresponding formula is

$$MS_{total} = MS_{regression} + MS_{residual}.$$

This can also be derived from the additivity of variances, since the total (observed) score is the sum of the predicted score and the error score, where the latter two are uncorrelated.

Similar decompositions are possible for the sum of squared deviations (sum of squares, SS):

$$SS_{total} = SS_{between} + SS_{within},$$
$$SS_{total} = SS_{regression} + SS_{residual}.$$

Formulae for the Variance

A formula often used for deriving the variance of a theoretical distribution is as follows:

$$Var(X) = E(X^2) - (E(X))^2.$$

This will be useful when it is possible to derive formulae for the expected value and for the expected value of the square.

This formula is also sometimes used in connection with the sample variance. While useful for hand calculations, it is not advised for computer calculations as it suffers from catastrophic cancellation if the two components of the equation are similar in magnitude and floating point arithmetic is used.

Calculation from the CDF

The population variance for a non-negative random variable can be expressed in terms of the cumulative distribution function F using

$$2\int_0^\infty u(1-F(u))du - \left(\int_0^\infty 1-F(u)du\right)^2.$$

This expression can be used to calculate the variance in situations where the CDF, but not the density, can be conveniently expressed.

Characteristic Property

The second moment of a random variable attains the minimum value when taken around the first moment (i.e., mean) of the random variable, i.e. $\text{argmin}_m E((X-m)^2) = E(X)$. Conversely, if a continuous function φ satisfies $\text{argmin}_m E(\varphi(X-m)) = E(X)$ for all random variables X, then it is necessarily of the form $\varphi(x) = ax^2 + b$, where $a > 0$. This also holds in the multidimensional case.

Units of Measurement

Unlike expected absolute deviation, the variance of a variable has units that are the square of the units of the variable itself. For example, a variable measured in meters will have a variance measured in meters squared. For this reason, describing data sets via their standard deviation or root mean square deviation is often preferred over using the variance. In the dice example the standard deviation is $\sqrt{2.9} \approx 1.7$, slightly larger than the expected absolute deviation of 1.5.

The standard deviation and the expected absolute deviation can both be used as an indicator of the "spread" of a distribution. The standard deviation is more amenable to algebraic manipulation than the expected absolute deviation, and, together with variance and its generalization covariance, is used frequently in theoretical statistics; however the expected absolute deviation tends to be more robust as it is less sensitive to outliers arising from measurement anomalies or an unduly heavy-tailed distribution.

Approximating the Variance of a Function

The delta method uses second-order Taylor expansions to approximate the variance of a function of one or more random variables. For example, the approximate variance of a function of one variable is given by

$$\text{Var}\left[f(X)\right] \approx \left(f'(E\left[X\right])\right)^2 \text{Var}\left[X\right]$$

provided that f is twice differentiable and that the mean and variance of X are finite.

Population Variance and Sample Variance

Real-world observations such as the measurements of yesterday's rain throughout the day typically cannot be complete sets of all possible observations that could be made. As such, the variance calculated from the finite set will in general not match the variance that would have been calculated from the full population of possible observations. This means that one estimates the mean and variance that would have been calculated from an omniscient set of observations by using an estimator equation. The estimator is a function of the sample of n observations drawn without observational bias from the whole population of potential observations. In this example that sample would be the set of actual measurements of yesterday's rainfall from available rain gauges within the geography of interest.

The simplest estimators for population mean and population variance are simply the mean and variance of the sample, the sample mean and (uncorrected) sample variance – these are consistent estimators (they converge to the correct value as the number of samples increases), but can be improved. Estimating the population variance by taking the sample's variance is close to optimal in general, but can be improved in two ways. Most simply, the sample variance is computed as an average of squared deviations about the (sample) mean, by dividing by n. However, using values other than n improves the estimator in various ways. Four common values for the denominator are n, $n-1$, $n+1$, and $n-1.5$: n is the simplest (population variance of the sample), $n-1$ eliminates bias, $n+1$ minimizes mean squared error for the normal distribution, and $n-1.5$ mostly eliminates bias in unbiased estimation of standard deviation for the normal distribution.

Firstly, if the omniscient mean is unknown (and is computed as the sample mean), then the sample variance is a biased estimator: it underestimates the variance by a factor of $(n-1)/n$; correcting by this factor (dividing by $n-1$ instead of n) is called Bessel's correction. The resulting estimator is unbiased, and is called the (corrected) sample variance or unbiased sample variance. For example, when $n=1$ the variance of a single observation about the sample mean (itself) is obviously zero regardless of the population variance. If the mean is determined in some other way than from the same samples used to estimate the variance then this bias does not arise and the variance can safely be estimated as that of the samples about the (independently known) mean.

Secondly, the sample variance does not generally minimize mean squared error between sample variance and population variance. Correcting for bias often makes this worse: one can always choose a scale factor that performs better than the corrected sample variance, though the optimal scale factor depends on the excess kurtosis of the population, and introduces bias. This always consists of scaling down the unbiased estimator (dividing by a number larger than $n-1$), and is a simple example of a shrinkage estimator: one "shrinks" the unbiased estimator towards zero. For the normal distribution, dividing by $n+1$ (instead of $n-1$ or n) minimizes mean squared error. The resulting estimator is biased, however, and is known as the biased sample variation.

Population Variance

In general, the *population variance* of a *finite* population of size N with values x_i is given by

$$\sigma^2 = \frac{1}{N}\sum_{i=1}^{N}(x_i - \mu)^2 = \frac{1}{N}\sum_{i=1}^{N}(x_i^2 - 2\mu x_i + \mu^2) = \left(\frac{1}{N}\sum_{i=1}^{N}x_i^2\right) - 2\mu\left(\frac{1}{N}\sum_{i=1}^{N}x_i\right) + \mu^2 = \left(\frac{1}{N}\sum_{i=1}^{N}x_i^2\right) - \mu^2$$

where the population mean is

$$\mu = \frac{1}{N}\sum_{i=1}^{N}x_i.$$

The population variance can also be computed using

$$\sigma^2 = \frac{1}{N^2}\sum_{i<j}(x_i - x_j)^2 = \frac{1}{2N^2}\sum_{i,j=1}^{N}(x_i - x_j)^2$$

This is true because

$$\frac{1}{2N^2}\sum_{i,j=1}^{N}(x_i - x_j)^2 = \frac{1}{2N^2}\sum_{i,j=1}^{N}(x_i^2 - 2x_i x_j + x_j^2)$$

$$= \frac{1}{2N}\sum_{j=1}^{N}\left(\frac{1}{N}\sum_{i=1}^{N}x_i^2\right) - \left(\frac{1}{N}\sum_{i=1}^{N}x_i\right)\left(\frac{1}{N}\sum_{j=1}^{N}x_j\right) + \frac{1}{2N}\sum_{i=1}^{N}\left(\frac{1}{N}\sum_{j=1}^{N}x_j^2\right)$$

$$= \frac{1}{2}\left(\sigma^2 + \mu^2\right) - \mu^2 + \frac{1}{2}\left(\sigma^2 + \mu^2\right) = \sigma^2$$

The population variance matches the variance of the generating probability distribution. In this sense, the concept of population can be extended to continuous random variables with infinite populations.

Sample Variance

In many practical situations, the true variance of a population is not known *a priori* and must be computed somehow. When dealing with extremely large populations, it is not possible to count every object in the population, so the computation must be performed on a sample of the population. Sample variance can also be applied to the estimation of the variance of a continuous distribution from a sample of that distribution.

We take a sample with replacement of n values $y_1, ..., y_n$ from the population, where $n < N$, and estimate the variance on the basis of this sample. Directly taking the variance of the sample data gives the average of the squared deviations:

$$\sigma_y^2 = \frac{1}{n}\sum_{i=1}^{n}(y_i - \bar{y})^2 = \left(\frac{1}{n}\sum_{i=1}^{n}y_i^2\right) - \bar{y}^2 = \frac{1}{n^2}\sum_{i<j}(y_i - y_j)^2.$$

Here, $\bar{y}$ denotes the sample mean:

$$\bar{y} = \frac{1}{n}\sum_{i=1}^{n}y_i.$$

Since the y_i are selected randomly, both $\bar{y}$ and σ_y^2 are random variables. Their expected values can be evaluated by averaging over the ensemble of all possible samples $\{y_i\}$ of size n from the population. For σ_y^2 this gives:

$$E[\sigma_y^2] = E\left[\frac{1}{n}\sum_{i=1}^{n}\left(y_i - \frac{1}{n}\sum_{j=1}^{n}y_j\right)^2\right]$$

$$= \frac{1}{n}\sum_{i=1}^{n}E\left[y_i^2 - \frac{2}{n}y_i\sum_{j=1}^{n}y_j + \frac{1}{n^2}\sum_{j=1}^{n}y_j\sum_{k=1}^{n}y_k\right]$$

$$= \frac{1}{n}\sum_{i=1}^{n}\left[\frac{n-2}{n}E[y_i^2] - \frac{2}{n}\sum_{j\neq i}E[y_iy_j] + \frac{1}{n^2}\sum_{j=1}^{n}\sum_{k\neq j}^{n}E[y_jy_k] + \frac{1}{n^2}\sum_{j=1}^{n}E[y_j^2]\right]$$

$$= \frac{1}{n}\sum_{i=1}^{n}\left[\frac{n-2}{n}(\sigma^2 + \mu^2) - \frac{2}{n}(n-1)\mu^2 + \frac{1}{n^2}n(n-1)\mu^2 + \frac{1}{n}(\sigma^2 + \mu^2)\right]$$

$$= \frac{n-1}{n}\sigma^2.$$

Hence σ_y^2 gives an estimate of the population variance that is biased by a factor of $\frac{n-1}{n}$. For this reason, σ_y^2 is referred to as the *biased sample variance*. Correcting for this bias yields the *unbiased sample variance*:

$$s^2 = \frac{n}{n-1}\sigma_y^2 = \frac{n}{n-1}\left(\frac{1}{n}\sum_{i=1}^{n}(y_i - \bar{y})^2\right) = \frac{1}{n-1}\sum_{i=1}^{n}(y_i - \bar{y})^2$$

Either estimator may be simply referred to as the *sample variance* when the version can be determined by context. The same proof is also applicable for samples taken from a continuous probability distribution.

The use of the term $n - 1$ is called Bessel's correction, and it is also used in sample covariance and the sample standard deviation (the square root of variance). The square root is a concave function and thus introduces negative bias (by Jensen's inequality), which depends on the distribution, and thus the corrected sample standard deviation (using Bessel's correction) is biased. The unbiased estimation of standard deviation is a technically involved problem, though for the normal distribution using the term $n - 1.5$ yields an almost unbiased estimator.

The unbiased sample variance is a U-statistic for the function $f(y_1, y_2) = (y_1 - y_2)^2/2$, meaning that it is obtained by averaging a 2-sample statistic over 2-element subsets of the population.

Distribution of the Sample Variance

Being a function of random variables, the sample variance is itself a random variable, and it is natural to study its distribution. In the case that y_i are independent observations from a normal distribution, Cochran's theorem shows that s^2 follows a scaled chi-squared distribution:

$$(n-1)\frac{s^2}{\sigma^2} \sim \chi^2_{n-1}.$$

As a direct consequence, it follows that

$$E(s^2) = E\left(\frac{\sigma^2}{n-1}\chi^2_{n-1}\right) = \sigma^2,$$

and

$$\mathrm{Var}[s^2] = \mathrm{Var}\left(\frac{\sigma^2}{n-1}\chi^2_{n-1}\right) = \frac{\sigma^4}{(n-1)^2}\mathrm{Var}\left(\chi^2_{n-1}\right) = \frac{2\sigma^4}{n-1}.$$

If the y_i are independent and identically distributed, but not necessarily normally distributed, then

$$E[s^2] = \sigma^2, \quad \mathrm{Var}[s^2] = \frac{\sigma^4}{n}\left((\kappa-1)+\frac{2}{n-1}\right) = \frac{1}{n}\left(\mu_4 - \frac{n-3}{n-1}\sigma^4\right),$$

where κ is the kurtosis of the distribution and μ_4 is the fourth central moment.

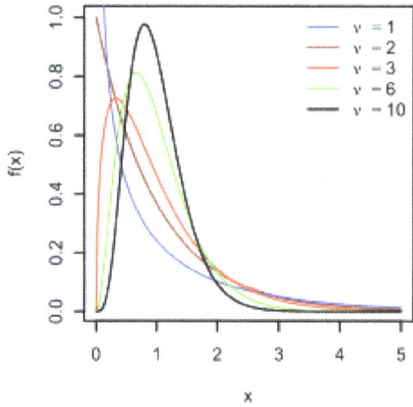

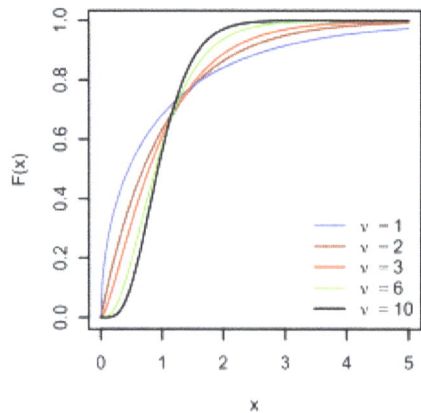

Distribution and cumulative distribution of s^2/σ^2, for various values of $v = n - 1$, when the y_i are independent normally distributed.

If the conditions of the law of large numbers hold for the squared observations, s^2 is a consistent estimator of σ^2. One can see indeed that the variance of the estimator tends asymptotically to zero. An asymptotically equivalent formula was given in Kenney and Keeping (1951:164), Rose and Smith (2002:264), and Weisstein (n.d.).

Samuelson's Inequality

Samuelson's inequality is a result that states bounds on the values that individual observations in a sample can take, given that the sample mean and (biased) variance have been calculated. Values must lie within the limits $\bar{y} \pm \sigma_y (n-1)^{1/2}$.

Relations with the Harmonic and Arithmetic Means

It has been shown that for a sample $\{y_i\}$ of real numbers,

$$\sigma_y^2 \leq 2y_{max}(A-H),$$

where y_{max} is the maximum of the sample, A is the arithmetic mean, H is the harmonic mean of the sample and σ_y^2 is the (biased) variance of the sample.

This bound has been improved, and it is known that variance is bounded by

$$\sigma_y^2 \leq \frac{y_{max}(A-H)(y_{max}-A)}{y_{max}-H},$$

$$\sigma_y^2 \geq \frac{y_{min}(A-H)(A-y_{min})}{H-y_{min}},$$

where y_{min} is the minimum of the sample.

Tests of Equality of Variances

Testing for the equality of two or more variances is difficult. The F test and chi square tests are both adversely affected by non-normality and are not recommended for this purpose.

Several non parametric tests have been proposed: these include the Barton–David–Ansari–Freund–Siegel–Tukey test, the Capon test, Mood test, the Klotz test and the Sukhatme test. The Sukhatme test applies to two variances and requires that both medians be known and equal to zero. The Mood, Klotz, Capon and Barton–David–Ansari–Freund–Siegel–Tukey tests also apply to two variances. They allow the median to be unknown but do require that the two medians are equal.

The Lehmann test is a parametric test of two variances. Of this test there are several variants known. Other tests of the equality of variances include the Box test, the Box–Anderson test and the Moses test.

Resampling methods, which include the bootstrap and the jackknife, may be used to test the equality of variances.

History

The term *variance* was first introduced by Ronald Fisher in his 1918 paper *The Correlation Between Relatives on the Supposition of Mendelian Inheritance*:

The great body of available statistics show us that the deviations of a human measurement from its mean follow very closely the Normal Law of Errors, and, therefore, that the variability may be uniformly measured by the standard deviation corresponding to the square root of the mean square error. When there are two independent causes of variability capable of producing in an otherwise uniform population distributions with standard deviations σ_1 and σ_2, it is found that the distribution, when both causes act together, has a standard deviation $\sqrt{\sigma_1^2 + \sigma_2^2}$. It is therefore desirable in analysing the causes of variability to deal with the square of the standard deviation as the measure of variability. We shall term this quantity the Variance...

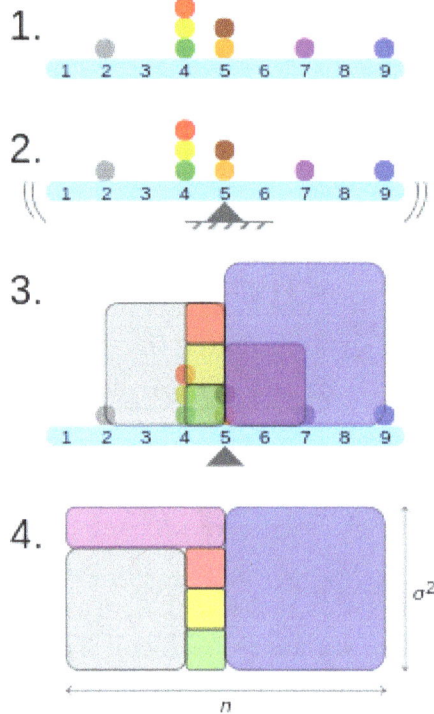

Geometric visualisation of the variance of an arbitrary distribution (2, 4, 4, 4, 5, 5, 7, 9):
1. A frequency distribution is constructed.
2. The centroid of the distribution gives its mean.
3. A square with sides equal to the difference of each value from the mean is formed for each value.
4. Arranging the squares into a rectangle with one side equal to the number of values, n, results in the other side being the distribution's variance, σ^2.

Moment of Inertia

The variance of a probability distribution is analogous to the moment of inertia in classical mechanics of a corresponding mass distribution along a line, with respect to rotation about its center of mass. It is because of this analogy that such things as the variance are called *moments* of probability distributions. The covariance matrix is related to the moment of inertia tensor for multivariate distributions. The moment of inertia of a cloud of n points with a covariance matrix of Σ is given by

$$I = n(\mathbf{1}_{3\times 3}\,\text{tr}(\Sigma) - \Sigma).$$

This difference between moment of inertia in physics and in statistics is clear for points that are gathered along a line. Suppose many points are close to the x axis and distributed along it. The covariance matrix might look like

$$\Sigma = \begin{bmatrix} 10 & 0 & 0 \\ 0 & 0.1 & 0 \\ 0 & 0 & 0.1 \end{bmatrix}.$$

That is, there is the most variance in the x direction. Physicists would consider this to have a low moment *about* the x axis so the moment-of-inertia tensor is

$$I = n \begin{bmatrix} 0.2 & 0 & 0 \\ 0 & 10.1 & 0 \\ 0 & 0 & 10.1 \end{bmatrix}.$$

Semivariance

The *semivariance* is calculated in the same manner as the variance but only those observations that fall below the mean are included in the calculation. It is sometimes described as a measure of downside risk in an investments context. For skewed distributions, the semivariance can provide additional information that a variance does not.

Generalizations

If x is a scalar complex-valued random variable, with values in $\mathbb{C}$, then its variance is $\text{E}\left[(x - \mu)(x - \mu)^*\right]$, where x^* is the complex conjugate of x This variance is a real scalar.

If X is a vector-valued random variable, with values in $\mathbb{R}^n$, and thought of as a column vector, then a natural generalization of variance is $\text{E}\left[(X - \mu)(X - \mu)^{\text{T}}\right]$, where $\mu = \text{E}(X)$ and X^{T} is the transpose of X and so is a row vector. The result is a positive semi-definite square matrix, commonly referred to as the variance-covariance matrix (or simply as the *covariance matrix*).

Another natural generalization of variance for such vector-valued random variables X which results in a scalar value rather than in a matrix, is obtained by interpreting the deviation between the random variable and its mean as the Euclidean distance. This results in $\text{E}\left[(X - \mu)^{\text{T}}(X - \mu)\right] = \text{tr}(C)$, which is the trace of the covariance matrix.

If X is a vector- and complex-valued random variable, with values in $\mathbb{C}^n$, then the covariance matrix is $\text{E}\left[(X - \mu)(X - \mu)^{\dagger}\right]$, where $X^{\dagger}$ is the conjugate transpose of X. This matrix is also positive semi-definite and square.

Using the assumption that y_i's are independently distributed, the variance of b_1 is

$$Var(b_1) = \sum_{i=1}^{n} k_i^2 Var(y_i) + \sum_i \sum_{j \neq i} k_i k_j Cov(y_i, y_j)$$

$$= \sigma^2 \frac{\sum_i (x_i - \bar{x})^2}{s_{xx}^2} \ (since \ y_i, ..., y_n \ are \ independent)$$

$$= \frac{\sigma^2 s_{xx}}{s_{xx}^2}$$

$$= \frac{\sigma^2}{s_{xx}}.$$

Similarly, the variance of b_0 is

$$Var(b_0) = Var(\bar{y}) + \bar{x}^2 Var(b_1) - 2\bar{x} Cov(\bar{y}, b_1).$$

First we find that

$$Cov(\bar{y}, b_1) = E\left[\{\bar{y} - E(\bar{y})\} \{b_1 - E(b_1)\} \right]$$

$$= E\left[\bar{\varepsilon}(\sum_i k_i y_j - \beta_1) \right]$$

$$= \frac{1}{n} E\left[(\sum_i \varepsilon_i)(\beta_0 \sum_i k_i + \beta_1 \sum_i k_i x_i + \sum_i k_i \varepsilon_i) - \beta_1 \sum_i \varepsilon_i \right]$$

$$= \frac{1}{n}\left[0 + 0 + 0 + 0 \right]$$

$$= 0$$

So

$$Var(b_0) = \sigma^2 \left(\frac{1}{n} + \frac{\bar{x}^2}{s_{xx}} \right).$$

Covariance

In probability theory and statistics, covariance is a measure of the joint variability of two random variables. If the greater values of one variable mainly correspond with the greater values of the other variable, and the same holds for the lesser values, i.e., the variables tend to show similar behavior, the covariance is positive. In the opposite case, when the greater values of one variable mainly correspond to the lesser values of the other, i.e., the variables tend to show opposite behavior, the covariance is negative. The sign of the covariance therefore shows the tendency in the linear relationship between the variables. The magnitude of the covariance is not easy to interpret. The normalized version of the covariance, the correlation coefficient, however, shows by its magnitude the strength of the linear relation.

A distinction must be made between (1) the covariance of two random variables, which is a population parameter that can be seen as a property of the joint probability distribution, and (2) the sample covariance, which in addition to serving as a descriptor of the sample, also serves as an estimated value of the population parameter.

Definition

The covariance between two jointly distributed real-valued random variables X and Y with finite second moments is defined as the expected product of their deviations from their individual expected values:

$$\text{cov}(X,Y) = \text{E}\big[(X - \text{E}[X])(Y - \text{E}[Y])\big],$$

where $\text{E}[X]$ is the expected value of X, also known as the mean of X. The covariance is also sometimes denoted "σ", in analogy to variance. By using the linearity property of expectations, this can be simplified to the expected value of their product minus the product of their expected values:

$$\begin{aligned}
\text{cov}(X,Y) &= \text{E}\Big[\big(X - \text{E}[X]\big)\big(Y - \text{E}[Y]\big)\Big] \\
&= \text{E}\Big[XY - X\,\text{E}[Y] - \text{E}[X]Y + \text{E}[X]\text{E}[Y]\Big] \\
&= \text{E}[XY] - \text{E}[X]\text{E}[Y] - \text{E}[X]\text{E}[Y] + \text{E}[X]\text{E}[Y] \\
&= \text{E}[XY] - \text{E}[X]\text{E}[Y].
\end{aligned}$$

However, when $\text{E}[XY] \approx \text{E}[X]\text{E}[Y]$, this last equation is prone to catastrophic cancellation when computed with floating point arithmetic and thus should be avoided in computer programs when the data has not been centered before. Numerically stable algorithms should be preferred in this case.

For random vectors $\mathbf{X} \in \mathbb{R}^m$ and $\mathbf{Y} \in \mathbb{R}^n$, the $m \times n$ cross covariance matrix (also known as dispersion matrix or variance–covariance matrix, or simply called covariance matrix) is equal to

$$\begin{aligned}
\text{cov}(\mathbf{X}, \mathbf{Y}) &= \text{E}\Big[(\mathbf{X} - \text{E}[\mathbf{X}])(\mathbf{Y} - \text{E}[\mathbf{Y}])^{\text{T}}\Big] \\
&= \text{E}\big[\mathbf{X}\mathbf{Y}^{\text{T}}\big] - \text{E}[\mathbf{X}]\text{E}[\mathbf{Y}]^{\text{T}},
\end{aligned}$$

where m^{T} is the transpose of the vector (or matrix) m.

The (i, j)-th element of this matrix is equal to the covariance $\text{cov}(X_i, Y_j)$ between the i-th scalar component of X and the j-th scalar component of Y. In particular, $\text{cov}(Y, X)$ is the transpose of $\text{cov}(X, Y)$.

For a vector $\mathbf{X} = \begin{bmatrix} X_1 & X_2 & \ldots & X_m \end{bmatrix}^T$ of m jointly distributed random variables with finite second moments, its covariance matrix is defined as

$$\Sigma(\mathbf{X}) = \text{cov}(\mathbf{X}, \mathbf{X}).$$

Random variables whose covariance is zero are called uncorrelated. Similarly, random vectors whose covariance matrix is zero in every entry outside the main diagonal are called uncorrelated.

The units of measurement of the covariance cov(X, Y) are those of X times those of Y. By contrast, correlation coefficients, which depend on the covariance, are a dimensionless measure of linear dependence. (In fact, correlation coefficients can simply be understood as a normalized version of covariance.)

Discrete Variables

If each variable has a finite set of equal-probability values, x_i and y_j respectively for $i = 1, \ldots, n$ and $j = 1, \ldots, k$, then the covariance can be equivalently written in terms of the means $E(X)$ and $E(Y)$ as

$$\text{cov}(X, Y) = \frac{1}{nk} \sum_{i=1}^{n} \sum_{j=1}^{k} (x_i - E(X))(y_j - E(Y)).$$

It can also be equivalently expressed, when $n = k$, without directly referring to the means, as

$$\text{cov}(X, Y) = \frac{1}{n^2} \sum_{i=1}^{n} \sum_{j=1}^{n} \frac{1}{2}(x_i - x_j) \cdot (y_i - y_j) = \frac{1}{n^2} \sum_{i} \sum_{j>i} (x_i - x_j) \cdot (y_i - y_j).$$

Discrete Random Variable Example

Suppose that X and Y have the following joint probability mass function:

			y		
	f(x,y)	1	2	3	$f_X(x)$
	1	0.25	0.25	0	0.5
x	2	0	0.25	0.25	0.5
	$f_Y(y)$	0.25	0.5	0.25	1

Then $\mu_X = \dfrac{3}{2}$, $\mu_Y = 2$, $\sigma_X = \dfrac{1}{2}$ and $\sigma_Y = \sqrt{\dfrac{1}{2}}$.

$$\text{cov}(X, Y) = \sigma_{XY} = \sum_{(x,y) \in S} (x - \mu_X)(y - \mu_Y) f(x, y)$$

$$= \left(1 - \frac{3}{2}\right)(1 - 2)\left(\frac{1}{4}\right) + \left(1 - \frac{3}{2}\right)(2 - 2)\left(\frac{1}{4}\right)$$

$$+\left(1-\frac{3}{2}\right)(3-2)(0)+\left(2-\frac{3}{2}\right)(1-2)(0)$$

$$+\left(2-\frac{3}{2}\right)(2-2)\left(\frac{1}{4}\right)+\left(2-\frac{3}{2}\right)(3-2)\left(\frac{1}{4}\right)$$

$$=\frac{1}{4}$$

Additional examples can be found here.

Properties

- The variance is a special case of the covariance in which the two variables are identical (that is, in which one variable always takes the same value as the other):

$$\mathrm{cov}(X,X)=\mathrm{var}(X)\equiv\sigma^2(X)\equiv\sigma_X^2.$$

- If X, Y, W, and V are real-valued random variables and a, b, c, d are constant ("constant" in this context means non-random), then the following facts are a consequence of the definition of covariance:

$$\sigma(X,a)\qquad\qquad=0$$
$$\sigma(X,X)\qquad\qquad=\sigma^2(X)$$
$$\sigma(X,Y)\qquad\qquad=\sigma(Y,X)$$
$$\sigma(aX,bY)\qquad\qquad=ab\sigma(X,Y)$$
$$\sigma(X+a,Y+b)\qquad=\sigma(X,Y)$$
$$\sigma(aX+bY,cW+dV)=ac\sigma(X,W)+ad\sigma(X,V)+bc\sigma(Y,W)+bd\sigma(Y,V)$$

For a sequence $X_1, ..., X_n$ of random variables, and constants $a_1, ..., a_n$, we have

$$\sigma^2\left(\sum_{i=1}^n a_i X_i\right)=\sum_{i=1}^n a_i^2\sigma^2(X_i)+2\sum_{i,j:i<j}a_i a_j\sigma(X_i,X_j)=\sum_{i,j}a_i a_j\sigma(X_i,X_j)$$

- A useful identity to compute the covariance between two random variables X,Y is the Hoeffding's Covariance Identity:

$$\mathrm{cov}(X,Y)=\iint_{\mathbb{R}\,\mathbb{R}}F_{(X,Y)}(x,y)-F_X(x)F_Y(y)dxdy$$

where $F_{(XY)}(x,y)$ is the joint distribution function of the random vector (X,Y) and $F_X(x),F_Y(y)$ are the marginals.

A More General Identity for Covariance Matrices

Let X be a random vector with covariance matrix $\Sigma(X)$, and let A be a matrix that can act on X. The covariance matrix of the matrix-vector product A X is:

$$\Sigma(\mathbf{AX})=\mathrm{E}[\mathbf{AXX}^\mathrm{T}\mathbf{A}^\mathrm{T}]-\mathrm{E}[\mathbf{AX}]\mathrm{E}[\mathbf{X}^\mathrm{T}\mathbf{A}^\mathrm{T}]=\mathbf{A}\Sigma(\mathbf{X})\mathbf{A}^\mathrm{T}.$$

This is a direct result of the linearity of expectation and is useful when applying a linear transformation, such as a whitening transformation, to a vector.

Uncorrelatedness and Independence

If X and Y are independent, then their covariance is zero. This follows because under independence,

$$E[XY] = E[X] \cdot E[Y].$$

The converse, however, is not generally true. For example, let X be uniformly distributed in $[-1, 1]$ and let $Y = X^2$. Clearly, X and Y are dependent, but

$$\begin{aligned}
\sigma(X,Y) &= \sigma(X, X^2) \\
&= E[X \cdot X^2] - E[X] \cdot E[X^2] \\
&= E\left[X^3\right] - E[X]E[X^2] \\
&= 0 - 0 \cdot E[X^2] \\
&= 0.
\end{aligned}$$

In this case, the relationship between Y and X is non-linear, while correlation and covariance are measures of linear dependence between two variables. This example shows that if two variables are uncorrelated, that does not in general imply that they are independent. However, if two variables are jointly normally distributed (but not if they are merely individually normally distributed), uncorrelatedness *does* imply independence.

Relationship to Inner Products

Many of the properties of covariance can be extracted elegantly by observing that it satisfies similar properties to those of an inner product:

1. bilinear: for constants a and b and random variables X, Y, Z, $\sigma(aX + bY, Z) = a \, \sigma(X, Z) + b \, \sigma(Y, Z)$;

2. symmetric: $\sigma(X, Y) = \sigma(Y, X)$;

3. positive semi-definite: $\sigma^2(X) = \sigma(X, X) \geq 0$ for all random variables X, and $\sigma(X, X) = 0$ implies that X is a constant random variable (K).

In fact these properties imply that the covariance defines an inner product over the quotient vector space obtained by taking the subspace of random variables with finite second moment and identifying any two that differ by a constant. (This identification turns the positive semi-definiteness above into positive definiteness.) That quotient vector space is isomorphic to the subspace of random variables with finite second moment and mean zero; on that subspace, the covariance is exactly the L_2 inner product of real-valued functions on the sample space.

As a result for random variables with finite variance, the inequality

$$|\sigma(X,Y)| \leq \sqrt{\sigma^2(X)\sigma^2(Y)}$$

holds via the Cauchy–Schwarz inequality.

Proof: If $\sigma^2(Y) = 0$, then it holds trivially. Otherwise, let random variable

$$Z = X - \frac{\sigma(X,Y)}{\sigma^2(Y)}Y.$$

Then we have

$$0 \leq \sigma^2(Z) = \sigma\left(X - \frac{\sigma(X,Y)}{\sigma^2(Y)}Y, X - \frac{\sigma(X,Y)}{\sigma^2(Y)}Y\right)$$

$$= \sigma^2(X) - \frac{(\sigma(X,Y))^2}{\sigma^2(Y)}.$$

Calculating the Sample Covariance

The sample covariance of N observations of K variables is the K-by-K matrix $\overline{\overline{q}} = \left[[q_{jk}]\right]$ with the entries

$$q_{jk} = \frac{1}{N-1}\sum_{i=1}^{N}\left(X_{ij} - \bar{X}_j\right)\left(X_{ik} - \bar{X}_k\right),$$

which is an estimate of the covariance between variable j and variable k.

The sample mean and the sample covariance matrix are unbiased estimates of the mean and the covariance matrix of the random vector $\mathbf{X}$, a row vector whose jth element ($j = 1, ..., K$) is one of the random variables. The reason the sample covariance matrix has $N-1$ in the denominator rather than N is essentially that the population mean $\mathrm{E}(X)$ is not known and is replaced by the sample mean $\bar{\mathbf{X}}$. If the population mean $\mathrm{E}(X)$ is known, the analogous unbiased estimate is given by

$$q_{jk} = \frac{1}{N}\sum_{i=1}^{N}\left(X_{ij} - \mathrm{E}(X_j)\right)\left(X_{ik} - \mathrm{E}(X_k)\right).$$

Comments

The covariance is sometimes called a measure of "linear dependence" between the two random variables. That does not mean the same thing as in the context of linear algebra. When the covariance is normalized, one obtains the correlation coefficient. From it, one can obtain the Pearson coefficient, which gives the goodness of the fit for the best possible linear function describing the relation between the variables. In this sense covariance is a linear gauge of dependence.

Applications

In genetics and Molecular Biology

Covariance is an important measure in biology. Certain sequences of DNA are conserved more than others among species, and thus to study secondary and tertiary structures of proteins, or of RNA structures, sequences are compared in closely related species. If sequence changes are found or no changes at all are found in noncoding RNA (such as microRNA), sequences are found to be necessary for common structural motifs, such as an RNA loop.

In Financial Economics

Covariances play a key role in financial economics, especially in portfolio theory and in the capital asset pricing model. Covariances among various assets' returns are used to determine, under certain assumptions, the relative amounts of different assets that investors should (in a normative analysis) or are predicted to (in a positive analysis) choose to hold in a context of diversification.

In Meteorological and Oceanographic Data Assimilation

The covariance matrix is important in estimating the initial conditions required for running weather forecast models. The 'forecast error covariance matrix' is typically constructed between perturbations around a mean state (either a climatological or ensemble mean). The 'observation error covariance matrix' is constructed to represent the magnitude of combined observational errors (on the diagonal) and the correlated errors between measurements (off the diagonal).

In Feature Extraction

The covariance matrix is used to capture the spectral variability of a signal.

The covariance between b_0 and b_1 is

$$Cov(b_0, b_1) = Cov(\bar{y}, b_1) - \bar{x}Var(b_1)$$

$$= -\frac{\bar{x}}{S_{xx}}\sigma^2.$$

It can further be shown that the ordinary least squares estimators b_0 and b_1 possess the minimum variance in the class of linear and unbiased estimators. So they are termed as the Best Linear Unbiased Estimators (BLUE). Such a property is known as the Gauss-Markov theorem which is discussed later in multiple linear regression model.

Residual Sum of Squares

In statistics, the residual sum of squares (RSS), also known as the sum of squared residuals (SSR) or the sum of squared errors of prediction (SSE), is the sum of the squares of residuals (deviations predicted from actual empirical values of data). It is a measure of the discrepancy between the data and an estimation model. A small RSS indicates a tight fit of the model to the data. It is used as an optimality criterion in parameter selection and model selection.

In general, total sum of squares = explained sum of squares + residual sum of squares. For a proof of this in the multivariate ordinary least squares (OLS) case.

One Explanatory Variable

In a model with a single explanatory variable, RSS is given by:

$$RSS = \sum_{i=1}^{n}(y_i - f(x_i))^2$$

where y_i is the ith value of the variable to be predicted, x_i is the ith value of the explanatory variable, and $f(x_i)$ is the predicted value of y_i (also termed $\widehat{y}_i$). In a standard linear simple regression model, $y_i = a + bx_i + \varepsilon_i$, where a and b are coefficients, y and x are the regressand and the regressor, respectively, and ε is the error term. The sum of squares of residuals is the sum of squares of estimates of ε_i; that is

$$RSS = \sum_{i=1}^{n}(\varepsilon_i)^2 = \sum_{i=1}^{n}(y_i - (\alpha + \beta x_i))^2$$

where α is the estimated value of the constant term α and β is the estimated value of the slope coefficient b.

Matrix Expression for the OLS Residual Sum of Squares

The general regression model with n observations and k explanators, the first of which is a constant unit vector whose coefficient is the regression intercept, is

$$y = X\beta + e$$

where y is an $n \times 1$ vector of dependent variable observations, each column of the $n \times k$ matrix X is a vector of observations on one of the k explanators, β is a $k \times 1$ vector of true coefficients, and e is an $n \times 1$ vector of the true underlying errors. The ordinary least squares estimator for β is

$$\hat{\beta} = (X^T X)^{-1} X^T y.$$

The residual vector $\hat{e}$ is $y - X\hat{\beta} = y - X(X^T X)^{-1} X^T y,$ so the residual sum of squares is:

$$RSS = \hat{e}^T \hat{e} = \| e \|^2,$$

(equivalent to the square-root of the norm of residuals); in full:

$$RSS = y^T y - y^T X(X^T X)^{-1} X^T y = y^T[I - X(X^T X)^{-1} X^T]y = y^T[I - H]y,$$

where H is the hat matrix, or the prediction matrix in linear regression.

The residual sum of squares is given as

$$SS_{res} = \sum_{i=1}^{n} \hat{\varepsilon}_i^2$$

$$= \sum_{i=1}^{n} (y_i - \hat{y}_i)^2$$

$$= \sum_{i=1}^{n} (y_i - b_0 - b_1 x_i)^2$$

$$= \sum_{i=1}^{n} \left[(y_i - \bar{y} + b_1 \bar{x} - b_1 x_i) \right]^2$$

$$= \sum_{i=1}^{n} \left[(y_i - \bar{y} - b_1 (x_i - \bar{x})) \right]^2$$

$$= \sum_{i=1}^{n} (y_i - \bar{y})^2 + b_1^2 \sum_{i=1}^{n} (x_i - \bar{x})^2 - 2b_1 \sum_{i=1}^{n} (x_i - \bar{x})(y_i - \bar{y})$$

$$= s_{yy} + b_1^2 s_{xx} - 2b_1^2 s_{xx}$$

$$= s_{yy} - b_1^2 s_{xx}$$

$$= s_{yy} - \left(\frac{s_{xy}}{s_{xx}} \right)^2 s_{xx}$$

$$= s_{yy} - \frac{s_{xy}^2}{s_{xx}}$$

$$= s_{yy} - b_1 s_{xy}.$$

Where

$$s_{yy} = \sum_{i=1}^{n} (y_i - \bar{y})^2, \bar{y} = \frac{1}{n} \sum_{i=1}^{n} y_i.$$

Estimation of σ^2

The estimator of σ^2 is obtained from residual sum of squares as follows. Assuming that Since y_i is normally distributed, so SS_{res} has a χ^2 distribution with (n - 2) degrees of freedom, so

$$\frac{SS_{res}}{\sigma^2} \sim \chi^2(n-2).$$

Thus using the result about the expectation of a chi-square random variable, we have

$$E(SS_{res}) = (n-2)\sigma^2.$$

Thus an unbiased estimator of is σ^2.

$$s^2 = \frac{SS_{res}}{(n-2)}$$

Note that SS_{res} has only (n - 2) degrees of freedom. The two degrees of freedom are lost due to estimation of b_0 and b_1. Since s^2 depends on the estimates b_0 and b_1, so it is a model dependent estimate of σ^2.

Estimate of Variances of b_0 and b_1

The estimators of variances of b_0 and b_1 are obtained by replacing σ^2 by $\hat{\sigma}^2 = s^2$ as follows:

$$\widehat{Var}(b_0) = s^2 \left(\frac{1}{n} + \frac{\bar{x}^2}{S_{xx}} \right)$$

and

$$\widehat{Var}(b_1) = \frac{s^2}{S_{xx}}$$

It is observed that since $\sum_{i=1}^{n}(y_i - \hat{y}_i) = 0$, so $\sum_{i=1}^{n} e_i = 0$. In the light of this property, e_i can be regarded as an estimate of unknown $\varepsilon_i (i = 1,...,n)$. This helps in verifying the different model assumptions on the basis of the given sample $(x_i, y_i), i = 1,...,n$

Further, note that

(i) $\sum_{i=1}^{n} x_i e_i = 0$,

(ii) $\sum_{i=1}^{n} \hat{y}_i e_i = 0$,

(iii) $\sum_{i=1}^{n} y_i = \sum_{i=1}^{n} \hat{y}_i$ and

(iv) the fitted line always passes through $(\bar{x}, \bar{y})$.

Centered Model

Sometimes it is useful to measure the independent variable around its mean. In such a case, model $y_i = \beta_0 + \beta_1 X_i + \varepsilon_i$ has a centered version as follows:

$$y_i = \beta_0 + \beta_1(x_i - \bar{x}) + \beta_1 \bar{x} + \varepsilon \quad (i = 1, 2,...,n)$$
$$= \beta_0^* + \beta_1(x_i - \bar{x}) + \varepsilon_i$$

where $\beta_0^* = \beta_0 + \beta_1 \bar{x}$. The sum of squares due to error is given by

$$S(\beta_0^*, \beta_1) = \sum_{i=1}^{n} \varepsilon_i^2 = \sum_{i=1}^{n} \left[y_i - \beta_0^* - \beta_1(x_i - \bar{x}) \right]^2 .$$

Now solving

$$\frac{\partial S(\beta_0^*, \beta_1)}{\partial \beta_0^*} = 0$$

$$\frac{\partial S(\beta_0^*, \beta_1)}{\partial \beta_1^*} = 0,$$

we get the direct regression least squares estimates of β_0^* and β_1 as

$$b_0^* = \bar{y}$$

and

$$b_1 = \frac{S_{xy}}{S_{xx}}$$

respectively.

Thus the form of the estimate of slope parameter β_1 remains same in usual and centered model whereas the form of the estimate of intercept term changes in the usual and centered models.

Further, the Hessian matrix of the second order partial derivatives of $S(\beta_0^*, \beta_1)$ with respect to β_0^* and β_1 is positive definite $\beta_0^* = b_1^*$ at and $\beta_1 = b_1$ which ensures that $S(\beta_0^*, \beta_1)$ is minimized at $\beta_0^* = b_0^*$ and $\beta_1 = b_1$.

Under the assumption that It follows that $E(\varepsilon_i) = 0, Var(\varepsilon_i) = \sigma^2$ and $Cov(\varepsilon_i \varepsilon_j) = 0$ for all $i \neq j = 1, 2, ..., n$. It follows that

$$E(b_0^*) = \beta_0^*, \quad E(b_1) = \beta_1,$$

$$Var(b_0^*) = \frac{\sigma^2}{n}, Var(b_1) = \frac{\sigma^2}{S_{xx}}.$$

In this case, the fitted model of $y_i = \beta_0^* + \beta_1(x_i - \bar{x}) + \varepsilon_i$ is

$$y = \bar{y} + b_1(x - \bar{x}),$$

and the predicted values are

$$\hat{y}_i = \bar{y} + b_1(x_i - \bar{x}) \quad (i = 1, ..., n).$$

Note that in centered model

$$Cov(b_0^*, b_1) = 0.$$

No Intercept Term Model

Sometimes in practice a model without an intercept term is used in those situations when $x_i = 0 \Rightarrow y_i = 0$ for all $i = 1, 2, ..., n$. A no-intercept model is

$$y_i = \beta_1 x_i + \varepsilon_i \, (i = 1, 2, ..., n).$$

For example, in analyzing the relationship between illumination of bulb (y) and electric current (X), the illumination of bulb is zero when current is zero.

Using the data $(x_i, y_i), i = 1, 2, ..., n.$ the direct regression least squares estimate of β_1 is obtained by minimizing

$$S(\beta_1) = \sum_{i=1}^{n} \varepsilon_i = \sum_{i=1}^{n} (y_i - \beta_1 x_i)^2$$

and solving

$$\frac{\partial S(\beta_1)}{\partial \beta_1} = 0$$

gives the estimator of β_1 as

$$b_1^* = \frac{\sum_{i=1}^{n} y_i x_i}{\sum_{i=1}^{n} x_i^2}$$

The second order partial derivative of $S(\beta_1)$ with respect to β_1 at $\beta_1 = b_1$ is positive which ensures that b1 minimizes $S(\beta_1)$.

Using the assumption that $E(\varepsilon_i) = 0, Var(\varepsilon_i) = \sigma^2$ and $Cov(\varepsilon_i \varepsilon_j) = 0$ for all $i \neq j = 1, 2, ..., n.$ the properties of b_1^* can be derived as follows:

$$E(b_1^*) = \frac{\sum_{i=1}^{n} x_i E(y_i)}{\sum_{i=1}^{n} x_i^2}$$

$$= \frac{\sum_{i=1}^{n} x_i^2 \beta_1}{\sum_{i=1}^{n} x_i^2}$$

$$= \beta_1.$$

This b_1^* is an unbiased estimator of β_1. The variance of b_1^* is obtained as follows:

$$Var(b_1^*) = \frac{\sum_{i=1}^{n} x_i^2 Var(y_i)}{\left(\sum_{i=1}^{n} x_i^2\right)^2}$$

$$= \sigma^2 \frac{\sum_{i=1}^{n} x_i^2}{\left(\sum_{i=1}^{n} x_i^2\right)^2}$$

$$= \frac{\sigma^2}{\sum_{i=1}^{n} x_i^2}$$

and an unbiased estimator of σ^2 is $\dfrac{\sum_{i=1}^{n} y_i^2 - b_1 \sum_{i=1}^{n} y_i x_i}{n-1}$.

Maximum Likelihood Estimation

In statistics, maximum likelihood estimation (MLE) is a method of estimating the parameters of a statistical model given observations, by finding the parameter values that maximize the likelihood of making the observations given the parameters. MLE can be seen as a special case of the maximum a posteriori estimation (MAP) that assumes a uniform prior distribution of the parameters, or as a variant of the MAP that ignores the prior and which therefore is un-regularized.

The method of maximum likelihood corresponds to many well-known estimation methods in statistics. For example, one may be interested in the heights of adult female penguins, but be unable to measure the height of every single penguin in a population due to cost or time constraints. Assuming that the heights are normally distributed with some unknown mean and variance, the mean and variance can be estimated with MLE while only knowing the heights of some sample of the overall population. MLE would accomplish this by taking the mean and variance as parameters and finding particular parametric values that make the observed results the most probable given the model.

In general, for a fixed set of data and underlying statistical model, the method of maximum likelihood selects the set of values of the model parameters that maximizes the likelihood function. Intuitively, this maximizes the "agreement" of the selected model with the observed data, and for discrete random variables it indeed maximizes the probability of the observed data under the resulting distribution. Maximum likelihood estimation gives a unified approach to estimation, which is well-defined in the case of the normal distribution and many other problems.

History

Ronald Fisher in 1913

Maximum-likelihood estimation was recommended, analyzed (with fruitless attempts at proofs) and widely popularized by Ronald Fisher between 1912 and 1922 (although it had been used earlier by Carl Friedrich Gauss, Pierre-Simon Laplace, Thorvald N. Thiele, and Francis Ysidro Edgeworth).

Maximum-likelihood estimation finally transcended heuristic justification in a proof published by Samuel S. Wilks in 1938, now called "Wilks' theorem". The theorem shows that the error in the logarithm of likelihood values for estimates from multiple independent samples is χ^2 distributed, which enables determination of a confidence region around any one estimate of the parameters. Ironically, the only difficult part of the proof depends on the expected value of the Fisher information matrix, which is provided by a theorem by Fisher. Wilks continued to improve on the generality of the theorem throughout his life, with his most general proof published in 1962.

Some of the theory behind maximum likelihood estimation was developed for Bayesian statistics. Reviews of the development of maximum likelihood estimation have been provided by a number of authors.

Principles

Suppose there is a sample x_1, x_2, ..., x_n of n independent and identically distributed observations, coming from a distribution with an unknown probability density function $f_0(\cdot)$. It is however surmised that the function f_0 belongs to a certain family of distributions $\{f(\cdot \,|\, \theta),\ \theta \in \Theta\}$ (where θ is a vector of parameters for this family), called the parametric model, so that $f_0 = f(\cdot \,|\, \theta_0)$. The value θ_0 is unknown and is referred to as the *true value* of the parameter vector. It is desirable to find an estimator $\hat{\theta}$ which would be as close to the true value θ_0 as possible. Either or both the observed variables x_i and the parameter θ can be vectors.

To use the method of maximum likelihood, one first specifies the joint density function for all observations. For an independent and identically distributed sample, this joint density function is

$$f(x_1, x_2, \ldots, x_n \,|\, \theta) = f(x_1 \,|\, \theta) \times f(x_2 \,|\, \theta) \times \cdots \times f(x_n \,|\, \theta).$$

Now we look at this function from a different perspective by considering the observed values x_1, x_2, ..., x_n to be fixed "parameters" of this function, whereas θ will be the function's variable and allowed to vary freely; this same function will be called the likelihood:

$$\mathcal{L}(\theta;x_1,\ldots,x_n) = f(x_1,x_2,\ldots,x_n \mid \theta) = \prod_{i=1}^{n} f(x_i \mid \theta).$$

Note that "$;$" denotes a separation between the two categories of input arguments: the parameters θ and the observations $x_1,\ldots,x_n$.

In practice the algebra is often more convenient when working with the natural logarithm of the likelihood function, called the log-likelihood:

$$\ln \mathcal{L}(\theta;x_1,\ldots,x_n) = \sum_{i=1}^{n} \ln f(x_i \mid \theta),$$

or the average log-likelihood:

$$\hat{\ell} = \frac{1}{n} \ln \mathcal{L}.$$

The hat over ℓ indicates that it is akin to some estimator. Indeed, $\hat{\ell}$ estimates the expected log-likelihood of a single observation in the model.

The method of maximum likelihood estimates θ_0 by finding a value of θ that maximizes $\hat{\ell}(\theta;x)$. This method of estimation defines a maximum likelihood estimator (MLE) of θ_0:

$$\{\hat{\theta}_{\text{mle}}\} \subseteq \{\underset{\theta \in \Theta}{\text{argmax }} \hat{\ell}(\theta;x_1,\ldots,x_n)\},$$

if a maximum exists. An MLE estimate is the same regardless of whether we maximize the likelihood or the log-likelihood function, since log is a monotonically increasing function.

For many models, a maximum likelihood estimator can be found as an explicit function of the observed data x_1, ..., x_n. For many other models, however, no closed-form solution to the maximization problem is known or available, and an MLE has to be found numerically using optimization methods. For some problems, there may be multiple estimates that maximize the likelihood. For other problems, no maximum likelihood estimate exists – either the log-likelihood function increases without ever reaching a supremum value, or the supremum does exist but is outside the bounds of Θ, the set of acceptable parameter values.

In the exposition above, it is assumed that the data are independent and identically distributed. The method can be applied however to a broader setting, as long as it is possible to write the joint density function $f(x_1, \ldots, x_n \mid \theta)$, and its parameter θ has a finite dimension which does not depend on the sample size n. In a simpler extension, an allowance can be made for data heterogeneity, so that the joint density is equal to $f_1(x_1 \mid \theta) \cdot f_2(x_2 \mid \theta) \cdot \cdots \cdot f_n(x_n \mid \theta)$. Put another way, we are now assuming that each observation x_i comes from a random variable that has its own distribution function f_i. In the more complicated case of time series models, the independence assumption may have to be dropped as well.

A maximum likelihood estimator coincides with the most probable Bayesian estimator given a uniform prior distribution on the parameters. Indeed, the maximum a posteriori estimate is the parameter θ that maximizes the probability of θ given the data, given by Bayes' theorem:

$$P(\theta \mid x_1, x_2, \ldots, x_n) = \frac{f(x_1, x_2, \ldots, x_n \mid \theta) P(\theta)}{P(x_1, x_2, \ldots, x_n)}$$

where $P(\theta)$ is the prior distribution for the parameter θ and where $P(x_1, x_2, \ldots, x_n)$ is the probability of the data averaged over all parameters. Since the denominator is independent of θ, the Bayesian estimator is obtained by maximizing $f(x_1, x_2, \ldots, x_n \mid \theta) P(\theta)$ with respect to θ. If we further assume that the prior $P(\theta)$ is a uniform distribution, the Bayesian estimator is obtained by maximizing the likelihood function $f(x_1, x_2, \ldots, x_n \mid \theta)$. Thus the Bayesian estimator coincides with the maximum likelihood estimator for a uniform prior distribution $P(\theta)$.

Properties

A maximum likelihood estimator is an extremum estimator obtained by maximizing, as a function of θ, the *objective function* (c.f., the loss function)

$$\hat{\ell}(\theta; x) = \frac{1}{n} \sum_{i=1}^{n} \ln f(x_i \mid \theta),$$

this being the sample analogue of the expected log-likelihood $\ell(\theta) = \mathrm{E}[\ln f(x_i \mid \theta)]$, where this expectation is taken with respect to the true density $f(\cdot \mid \theta_0)$.

Maximum-likelihood estimators have no optimum properties for finite samples, in the sense that (when evaluated on finite samples) other estimators may have greater concentration around the true parameter-value. However, like other estimation methods, maximum likelihood estimation possesses a number of attractive limiting properties: As the sample size increases to infinity, sequences of maximum likelihood estimators have these properties:

- Consistency: the sequence of MLEs converges in probability to the value being estimated.

- Asymptotic normality: as the sample size increases, the distribution of the MLE tends to the Gaussian distribution with mean θ and covariance matrix equal to the inverse of the Fisher information matrix.

- Efficiency, i.e., it achieves the Cramér–Rao lower bound when the sample size tends to infinity. This means that no consistent estimator has lower asymptotic mean squared error than the MLE (or other estimators attaining this bound).

- Second-order efficiency after correction for bias.

Consistency

Under the conditions outlined below, the maximum likelihood estimator is consistent. The consistency means that having a sufficiently large number of observations n, it is possible to find the value of θ_0 with arbitrary precision. In mathematical terms this means that as n goes to infinity the estimator converges in probability to its true value:

$$\hat{\theta}_{\text{mle}} \xrightarrow{\ p\ } \theta_0. \tag{1}$$

Under slightly stronger conditions, the estimator converges almost surely (or *strongly*) to:

$$\hat{\theta}_{\text{mle}} \xrightarrow{\ a.s\ } \theta_0. \tag{2}$$

To establish consistency, the following conditions are sufficient:

1. Identification of the model:

$$\theta \neq \theta_0 \quad \Leftrightarrow \quad f(\cdot \mid \theta) \neq f(\cdot \mid \theta_0).$$

In other words, different parameter values θ correspond to different distributions within the model. If this condition did not hold, there would be some value θ_1 such that θ_1 and θ_0 generate an identical distribution of the observable data. Then we would not be able to distinguish between these two parameters even with an infinite amount of data—these parameters would have been observationally equivalent.

The identification condition is absolutely necessary for the ML estimator to be consistent. When this condition holds, the limiting likelihood function $\ell(\theta \mid \cdot)$ has unique global maximum at θ_0.

2. Compactness: the parameter space Θ of the model is compact.

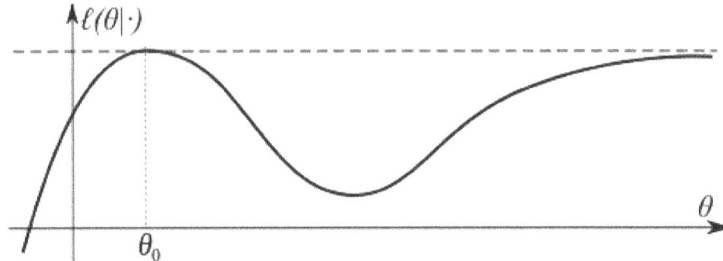

The identification condition establishes that the log-likelihood has a unique global maximum. Compactness implies that the likelihood cannot approach the maximum value arbitrarily close at some other point (as demonstrated for example in the picture on the right).

Compactness is only a sufficient condition and not a necessary condition. Compactness can be replaced by some other conditions, such as:

o   both concavity of the log-likelihood function and compactness of some (nonempty) upper level sets of the log-likelihood function, or

o   existence of a compact neighborhood N of θ_0 such that outside of N the log-likelihood function is less than the maximum by at least some $\varepsilon > 0$.

3. Continuity: the function $\ln f(x \mid \theta)$ is continuous in θ for almost all values of x:

$$\Pr\left[\ \ln f(x \mid \theta) \in \mathbb{C}^0(\Theta)\ \right] = 1.$$

The continuity here can be replaced with a slightly weaker condition of upper semi-continuity.

4. Dominance: there exists $D(x)$ integrable with respect to the distribution $f(x|\theta_0)$ such that

$$\left| \ln f(x|\theta) \right| < D(x) \quad \text{for all } \theta \in \Theta.$$

By the uniform law of large numbers, the dominance condition together with continuity establish the uniform convergence in probability of the log-likelihood:

$$\sup_{\theta \in \Theta} \left| \hat{\ell}(\theta|x) - \ell(\theta) \right| \xrightarrow{\ p\ } 0.$$

The dominance condition can be employed in the case of i.i.d. observations. In the non-i.i.d. case the uniform convergence in probability can be checked by showing that the sequence $\hat{\ell}(\theta|x)$ is stochastically equicontinuous.

If one wants to demonstrate that the ML estimator $\hat{\theta}$ converges to θ_0 almost surely, then a stronger condition of uniform convergence almost surely has to be imposed:

$$\sup_{\theta \in \Theta} \left\| \hat{\ell}(x|\theta) - \ell(\theta) \right\| \xrightarrow{\ a.s.\ } 0.$$

Asymptotic Normality

In a wide range of situations, maximum likelihood parameter estimates exhibit asymptotic normality – that is, they are equal to the true parameters plus a random error that is approximately normal (given sufficient data), and the error's variance decays as 1/n. For this property to hold, it is necessary that the estimator does not suffer from the following issues:

Estimate on Boundary

Sometimes the maximum likelihood estimate lies on the boundary of the set of possible parameters, or (if the boundary is not, strictly speaking, allowed) the likelihood gets larger and larger as the parameter approaches the boundary. Standard asymptotic theory needs the assumption that the true parameter value lies away from the boundary. If we have enough data, the maximum likelihood estimate will keep away from the boundary too. But with smaller samples, the estimate can lie on the boundary. In such cases, the asymptotic theory clearly does not give a practically useful approximation. Examples here would be variance-component models, where each component of variance, σ^2, must satisfy the constraint $\sigma^2 \geq 0$.

Data Boundary Parameter-dependent

For the theory to apply in a simple way, the set of data values which has positive probability (or positive probability density) should not depend on the unknown parameter. A simple example where such parameter-dependence does hold is the case of estimating θ from a set of independent identically distributed observations when the common distribution is uniform on the range $(0,\theta)$. For estimation purposes the relevant range of θ is such that θ cannot be less than the largest observation. Because the interval $(0,\theta)$ is not compact, there exists no maximum for the likelihood function: For any estimate of theta, there exists a greater estimate that also has greater likelihood. In contrast, the interval $[0,\theta]$ includes the end-point θ and is compact, in which case the maximum likelihood estimator exists. However, in this case, the maximum likelihood estimator is biased. Asymptotically, this maximum likelihood estimator is not normally distributed.

Nuisance Parameters

For maximum likelihood estimations, a model may have a number of nuisance parameters. For the asymptotic behaviour outlined to hold, the number of nuisance parameters should not increase with the number of observations (the sample size). A well-known example of this case is where observations occur as pairs, where the observations in each pair have a different (unknown) mean but otherwise the observations are independent and normally distributed with a common variance. Here for $2N$ observations, there are $N + 1$ parameters. It is well known that the maximum likelihood estimate for the variance does not converge to the true value of the variance.

Increasing Information

For the asymptotics to hold in cases where the assumption of independent identically distributed observations does not hold, a basic requirement is that the amount of information in the data increases indefinitely as the sample size increases. Such a requirement may not be met if either there is too much dependence in the data (for example, if new observations are essentially identical to existing observations), or if new independent observations are subject to an increasing observation error.

Some regularity conditions which ensure this behavior are:

1. The first and second derivatives of the log-likelihood function exist (are "well defined").

2. The Fisher information matrix is non-singular.

3. The Fisher information matrix is continuous as a function of the parameters, θ.

4. The maximum likelihood estimator is consistent.

Suppose that conditions for consistency of maximum likelihood estimator are satisfied, and

1. $\theta_0 \in$ interior(Θ);

2. $f(x \mid \theta) > 0$ and is twice continuously differentiable in Θ in some neighborhood N of θ_0;

3. $\int \sup_{\theta \in N} ||\nabla_\theta f(x \mid \theta)|| dx < \infty$, and $\int \sup_{\theta \in N} ||\nabla_{\theta\theta} f(x \mid \theta)|| dx < \infty$;

4. $I = E[\nabla_\theta \ln f(x \mid \theta_0) \nabla_\theta \ln f(x \mid \theta_0)']$ exists and is nonsingular;

5. $E[\sup_{\theta \in N} ||\nabla_{\theta\theta} \ln f(x \mid \theta)||] < \infty$.

Then the maximum likelihood estimator has asymptotically normal distribution:

$$\sqrt{n}\left(\hat{\theta}_{\text{mle}} - \theta_0\right) \overset{d}{\longrightarrow} \mathcal{N}\left(0, I^{-1}\right).$$

Sketch of Proof

Since the log-likelihood function is differentiable, and θ_0 lies in the interior of the parameter set Θ, in the maximum the first-order condition will be satisfied:

$$\nabla_\theta \hat{\ell}(\hat{\theta} \mid x) = \frac{1}{n} \sum_{i=1}^{n} \nabla_\theta \ln f(x_i \mid \hat{\theta}) = 0.$$

When the log-likelihood is twice differentiable, this expression can be expanded into a Taylor series around the point $\theta = \theta_0$:

$$0 = \frac{1}{n} \sum_{i=1}^{n} \nabla_\theta \ln f(x_i \mid \theta_0) + \left[\frac{1}{n} \sum_{i=1}^{n} \nabla_{\theta\theta} \ln f(x_i \mid \tilde{\theta}) \right](\hat{\theta} - \theta_0),$$

where $\tilde{\theta}$ is some point intermediate between θ_0 and $\hat{\theta}$. From this expression we can derive that

$$\sqrt{n}(\hat{\theta} - \theta_0) = \left[-\frac{1}{n} \sum_{i=1}^{n} \nabla_{\theta\theta} \ln f(x_i \mid \tilde{\theta}) \right]^{-1} \frac{1}{\sqrt{n}} \sum_{i=1}^{n} \nabla_\theta \ln f(x_i \mid \theta_0)$$

Here the expression in square brackets converges in probability to $H = \mathbb{E}\left[-\nabla_{\theta\theta} \ln f(x \mid \theta_0) \right]$ by the law of large numbers. The continuous mapping theorem ensures that the inverse of this expression also converges in probability, to H^{-1}. The second sum, by the central limit theorem, converges in distribution to a multivariate normal with mean zero and variance matrix equal to the Fisher information I. Thus, applying Slutsky's theorem to the whole expression, we obtain that

$$\sqrt{n}(\hat{\theta} - \theta_0) \xrightarrow{d} \mathcal{N}\left(0, H^{-1} I H^{-1} \right).$$

Finally, the information equality guarantees that when the model is correctly specified, matrix H will be equal to the Fisher information I, so that the variance expression simplifies to just I^{-1}.

Functional Invariance

The maximum likelihood estimator selects the parameter value which gives the observed data the largest possible probability (or probability density, in the continuous case). If the parameter consists of a number of components, then we define their separate maximum likelihood estimators, as the corresponding component of the MLE of the complete parameter. Consistent with this, if $\hat{\theta}$ is the MLE for θ, and if $g(\theta)$ is any transformation of θ, then the MLE for $\alpha = g(\theta)$ is by definition

$$\hat{\alpha} = g(\hat{\theta}).$$

It maximizes the so-called profile likelihood:

$$\bar{L}(\alpha) = \sup_{\theta : \alpha = g(\theta)} L(\theta).$$

The MLE is also invariant with respect to certain transformations of the data. If $Y = g(X)$ where g is one to one and does not depend on the parameters to be estimated, then the density functions satisfy

$$f_Y(y) = \frac{f_X(x)}{|g'(x)|}$$

and hence the likelihood functions for X and Y differ only by a factor that does not depend on the model parameters.

For example, the MLE parameters of the log-normal distribution are the same as those of the normal distribution fitted to the logarithm of the data.

Higher-order Properties

The standard asymptotics tells that the maximum likelihood estimator is $\sqrt{n}$-consistent and asymptotically efficient, meaning that it reaches the Cramér–Rao bound:

$$\sqrt{n}(\hat{\theta}_{\mathrm{mle}} - \theta_0) \xrightarrow{\ d\ } \mathcal{N}(0, I^{-1}),$$

where I is the Fisher information matrix:

$$I_{jk} = \mathrm{E}_X\left[-\frac{\partial^2 \ln f_{\theta_0}(X_t)}{\partial\theta_j \partial\theta_k} \right].$$

In particular, it means that the bias of the maximum likelihood estimator is equal to zero up to the order $n^{-1/2}$. However, when we consider the higher-order terms in the expansion of the distribution of this estimator, it turns out that θ_{mle} has bias of order n^{-1}. This bias is equal to (componentwise)

$$b_s \equiv \mathrm{E}[(\hat{\theta}_{\mathrm{mle}} - \theta_0)_s] = \frac{1}{n} \cdot I^{si} I^{jk} \left(\tfrac{1}{2} K_{ijk} + J_{j,ik} \right)$$

where Einstein's summation convention over the repeating indices has been adopted; I^{jk} denotes the j,k-th component of the *inverse* Fisher information matrix I^{-1}, and

$$\tfrac{1}{2} K_{ijk} + J_{j,ik} = \mathrm{E}_X\left[\frac{1}{2}\frac{\partial^3 \ln f_{\theta_0}(X_t)}{\partial\theta_i \partial\theta_j \partial\theta_k} + \frac{\partial \ln f_{\theta_0}(X_t)}{\partial\theta_j} \frac{\partial^2 \ln f_{\theta_0}(X_t)}{\partial\theta_i \partial\theta_k} \right].$$

Using these formulas it is possible to estimate the second-order bias of the maximum likelihood estimator, and *correct* for that bias by subtracting it:

$$\hat{\theta}^*_{\mathrm{mle}} = \hat{\theta}_{\mathrm{mle}} - \hat{b}.$$

This estimator is unbiased up to the terms of order n^{-1}, and is called the bias-corrected maximum likelihood estimator.

This bias-corrected estimator is *second-order efficient* (at least within the curved exponential family), meaning that it has minimal mean squared error among all second-order bias-corrected estimators, up to the terms of the order n^{-2}. It is possible to continue this process, that is to derive the third-order bias-correction term, and so on. However, as was shown by Kano (1996), the maximum likelihood estimator is not third-order efficient.

Examples

Discrete Uniform Distribution

Consider a case where n tickets numbered from 1 to n are placed in a box and one is selected at random; thus, the sample size is 1. If n is unknown, then the maximum likelihood estimator $\hat{n}$ of n is the number m on the drawn ticket. (The likelihood is 0 for $n < m$, $1/n$ for $n \geq m$, and this is greatest when $n = m$. Note that the maximum likelihood estimate of n occurs at the lower extreme of possible values $\{m, m + 1, ...\}$, rather than somewhere in the "middle" of the range of possible values, which would result in less bias.) The expected value of the number m on the drawn ticket, and therefore the expected value of , is $(n + 1)/2$. As a result, with a sample size of 1, the maximum likelihood estimator for n will systematically underestimate n by $(n - 1)/2$.

Discrete Distribution, Finite Parameter Space

Suppose one wishes to determine just how biased an unfair coin is. Call the probability of tossing a HEAD p. The goal then becomes to determine p.

Suppose the coin is tossed 80 times: i.e., the sample might be something like x_1 = H, x_2 = T, ..., x_{80} = T, and the count of the number of HEADS "H" is observed.

The probability of tossing TAILS is $1 - p$ (so here p is θ above). Suppose the outcome is 49 HEADS and 31 TAILS, and suppose the coin was taken from a box containing three coins: one which gives HEADS with probability $p = 1/3$, one which gives HEADS with probability $p = 1/2$ and another which gives HEADS with probability $p = 2/3$. The coins have lost their labels, so which one it was is unknown. Using maximum likelihood estimation the coin that has the largest likelihood can be found, given the data that were observed. By using the probability mass function of the binomial distribution with sample size equal to 80, number successes equal to 49 but different values of p (the "probability of success"), the likelihood function (defined below) takes one of three values:

$$\Pr(H = 49 \mid p = 1/3) = \binom{80}{49}(1/3)^{49}(1-1/3)^{31} \approx 0.000,$$

$$\Pr(H = 49 \mid p = 1/2) = \binom{80}{49}(1/2)^{49}(1-1/2)^{31} \approx 0.012,$$

$$\Pr(H = 49 \mid p = 2/3) = \binom{80}{49}(2/3)^{49}(1-2/3)^{31} \approx 0.054.$$

The likelihood is maximized when $p = 2/3$, and so this is the *maximum likelihood estimate* for p.

Discrete Distribution, Continuous Parameter Space

Now suppose that there was only one coin but its p could have been any value $0 \leq p \leq 1$. The likelihood function to be maximised is

$$L(p) = f_D(H = 49 \mid p) = \binom{80}{49} p^{49}(1-p)^{31},$$

and the maximisation is over all possible values $0 \le p \le 1$.

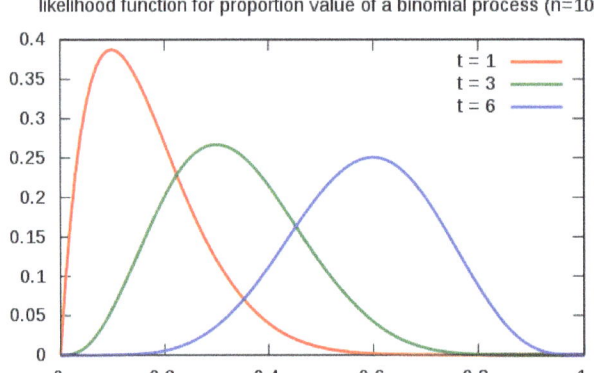

likelihood function for proportion value of a binomial process ($n = 10$)

One way to maximize this function is by differentiating with respect to p and setting to zero:

$$0 = \frac{\partial}{\partial p}\left(\binom{80}{49} p^{49}(1-p)^{31} \right),$$

$$0 = 49 p^{48}(1-p)^{31} - 31 p^{49}(1-p)^{30}$$

$$= p^{48}(1-p)^{30}\left[49(1-p) - 31p \right]$$

$$= p^{48}(1-p)^{30}\left[49 - 80p \right]$$

which has solutions $p = 0$, $p = 1$, and $p = 49/80$. The solution which maximizes the likelihood is clearly $p = 49/80$ (since $p = 0$ and $p = 1$ result in a likelihood of zero). Thus the *maximum likelihood estimator* for p is $49/80$.

This result is easily generalized by substituting a letter such as t in the place of 49 to represent the observed number of 'successes' of our Bernoulli trials, and a letter such as n in the place of 80 to represent the number of Bernoulli trials. Exactly the same calculation yields the *maximum likelihood estimator* t / n for any sequence of n Bernoulli trials resulting in t 'successes'.

Continuous Distribution, Continuous Parameter Space

For the normal distribution $\mathcal{N}(\mu, \sigma^2)$ which has probability density function

$$f(x \mid \mu, \sigma^2) = \frac{1}{\sqrt{2\pi\sigma^2}} \exp\left(-\frac{(x-\mu)^2}{2\sigma^2} \right),$$

the corresponding probability density function for a sample of n independent identically distributed normal random variables (the likelihood) is

$$f(x_1,\ldots,x_n \mid \mu,\sigma^2) = \prod_{i=1}^{n} f(x_i \mid \mu,\sigma^2) = \left(\frac{1}{2\pi\sigma^2}\right)^{n/2} \exp\left(-\frac{\sum_{i=1}^{n}(x_i - \mu)^2}{2\sigma^2}\right),$$

or more conveniently:

$$f(x_1,\ldots,x_n \mid \mu,\sigma^2) = \left(\frac{1}{2\pi\sigma^2}\right)^{n/2} \exp\left(-\frac{\sum_{i=1}^{n}(x_i - \overline{x})^2 + n(\overline{x} - \mu)^2}{2\sigma^2}\right),$$

where $\overline{x}$ is the sample mean.

This family of distributions has two parameters: $\theta = (\mu, \sigma)$, so we maximize the likelihood, $\mathcal{L}(\mu,\sigma) = f(x_1,\ldots,x_n \mid \mu,\sigma)$, over both parameters simultaneously, or if possible, individually.

Since the logarithm function itself is a continuous strictly increasing function over the range of the likelihood, the values which maximize the likelihood will also maximize its logarithm (The likelihood's logarithm is not strictly increasing). This log likelihood can be written as follows:

$$\log(\mathcal{L}(\mu,\sigma)) = (-n/2)\log(2\pi\sigma^2) - \frac{1}{2\sigma^2}\sum_{i=1}^{n}(x_i - \mu)^2$$

(Note: the log-likelihood is closely related to information entropy and Fisher information.)

We now compute the derivatives of this log likelihood as follows.

$$0 = \frac{\partial}{\partial \mu}\log(\mathcal{L}(\mu,\sigma)) = 0 - \frac{-2n(\overline{x} - \mu)}{2\sigma^2}.$$

This is solved by

$$\hat{\mu} = \overline{x} = \sum_{i=1}^{n}\frac{x_i}{n}.$$

This is indeed the maximum of the function since it is the only turning point in μ and the second derivative is strictly less than zero. Its expectation value is equal to the parameter μ of the given distribution,

$$E\left[\hat{\mu}\right] = \mu,$$

which means that the maximum likelihood estimator $\hat{\mu}$ is unbiased.

Similarly we differentiate the log likelihood with respect to σ and equate to zero:

$$0 = \frac{\partial}{\partial\sigma}\log\left(\left(\frac{1}{2\pi\sigma^2}\right)^{n/2}\exp\left(-\frac{\sum_{i=1}^{n}(x_i-\bar{x})^2 + n(\bar{x}-\mu)^2}{2\sigma^2}\right)\right)$$

$$= \frac{\partial}{\partial\sigma}\left(\frac{n}{2}\log\left(\frac{1}{2\pi\sigma^2}\right) - \frac{\sum_{i=1}^{n}(x_i-\bar{x})^2 + n(\bar{x}-\mu)^2}{2\sigma^2}\right)$$

$$= -\frac{n}{\sigma} + \frac{\sum_{i=1}^{n}(x_i-\bar{x})^2 + n(\bar{x}-\mu)^2}{\sigma^3}$$

which is solved by

$$\hat{\sigma}^2 = \frac{1}{n}\sum_{i=1}^{n}(x_i-\mu)^2.$$

Inserting the estimate $\mu = \hat{\mu}$ we obtain

$$\hat{\sigma}^2 = \frac{1}{n}\sum_{i=1}^{n}(x_i-\bar{x})^2 = \frac{1}{n}\sum_{i=1}^{n}x_i^2 - \frac{1}{n^2}\sum_{i=1}^{n}\sum_{j=1}^{n}x_i x_j.$$

To calculate its expected value, it is convenient to rewrite the expression in terms of zero-mean random variables (statistical error) $\delta_i \equiv \mu - x_i$. Expressing the estimate in these variables yields

$$\hat{\sigma}^2 = \frac{1}{n}\sum_{i=1}^{n}(\mu-\delta_i)^2 - \frac{1}{n^2}\sum_{i=1}^{n}\sum_{j=1}^{n}(\mu-\delta_i)(\mu-\delta_j).$$

Simplifying the expression above, utilizing the facts that $E[\delta_i] = 0$ and $E[\delta_i^2] = \sigma^2$, allows us to obtain

$$E[\hat{\sigma}^2] = \frac{n-1}{n}\sigma^2.$$

This means that the estimator $\hat{\sigma}$ is biased. However, $\hat{\sigma}$ is consistent.

Formally we say that the *maximum likelihood estimator* for $\theta = (\mu, \sigma^2)$ is:

$$\hat{\theta} = \left(\hat{\mu}, \hat{\sigma}^2\right).$$

In this case the MLEs could be obtained individually. In general this may not be the case, and the MLEs would have to be obtained simultaneously.

The normal log likelihood at its maximum takes a particularly simple form:

$$\log(\mathcal{L}(\hat{\mu}, \hat{\sigma})) = \frac{-n}{2}(\log(2\pi\hat{\sigma}^2) + 1)$$

This maximum log likelihood can be shown to be the same for more general least squares, even for non-linear least squares. This is often used in determining likelihood-based approximate confidence intervals and confidence regions, which are generally more accurate than those using the asymptotic normality discussed above.

Non-independent Variables

It may be the case that variables are correlated, that is, not independent. Two random variables X and Y are independent only if their joint probability density function is the product of the individual probability density functions, i.e.

$$f(x, y) = f(x)f(y)$$

Suppose one constructs an order-n Gaussian vector out of random variables $(x_1, \ldots, x_n)$, where each variable has means given by $(\mu_1, \ldots, \mu_n)$. Furthermore, let the covariance matrix be denoted by Σ.

The joint probability density function of these n random variables is then given by:

$$f(x_1, \ldots, x_n) = \frac{1}{(2\pi)^{n/2}\sqrt{\det(\Sigma)}} \exp\left(-\frac{1}{2}[x_1 - \mu_1, \ldots, x_n - \mu_n]\Sigma^{-1}[x_1 - \mu_1, \ldots, x_n - \mu_n]^T\right)$$

In the two variable case, the joint probability density function is given by:

$$f(x, y) = \frac{1}{2\pi\sigma_x\sigma_y\sqrt{1-\rho^2}} \exp\left[-\frac{1}{2(1-\rho^2)}\left(\frac{(x-\mu_x)^2}{\sigma_x^2} - \frac{2\rho(x-\mu_x)(y-\mu_y)}{\sigma_x\sigma_y} + \frac{(y-\mu_y)^2}{\sigma_y^2}\right)\right]$$

In this and other cases where a joint density function exists, the likelihood function is defined as above.

Iterative Procedures

Consider problems where both states x_i and parameters such as σ^2 require to be estimated. Iterative procedures such as Expectation-maximization algorithms may be used to solve joint state-parameter estimation problems.

For example, suppose that n samples of state estimates $\hat{x}_i$ together with a sample mean $\bar{x}$ have been calculated by either a minimum-variance Kalman filter or a minimum-variance smoother using a previous variance estimate $\hat{\sigma}^2$. Then the next variance iterate may be obtained from the maximum likelihood estimate calculation

$$\hat{\sigma}^2 = \frac{1}{n}\sum_{i=1}^{n}(\hat{x}_i - \bar{x})^2.$$

The convergence of MLEs within filtering and smoothing EM algorithms has been studied in the literature.

Applications

Maximum likelihood estimation is used for a wide range of statistical models, including:

- linear models and generalized linear models;

- exploratory and confirmatory factor analysis;

- structural equation modeling;

- many situations in the context of hypothesis testing and confidence intervals;

- discrete choice models;

- signal detection (filtering).

These uses arise across applications in widespread set of fields, including:

- communication systems;

- psychometrics;

- econometrics;

- time-delay of arrival (TDOA) in acoustic or electromagnetic detection;

- data modeling in nuclear and particle physics;

- magnetic resonance imaging;

- computational phylogenetics;

- origin/destination and path-choice modeling in transport networks;

- geographical satellite-image classification;

- power system state estimation.

We assume that ε_i 's $(i=1,2,...,n)$ are independent and identically distributed following a normal distribution $N(0,\sigma^2)$.

Now we use the method of maximum likelihood to estimate the parameters of the linear regression model

$$y_i = \beta_0 + \beta_1 x_i + \varepsilon_i (i=1,2,...,n),$$

the observations $y_i(i=1,2,...,n)$ are independently distributed with $N(\beta_0+\beta_1 x_i,\sigma^2)$ for all $i=1,2,...,n$. The likelihood function of the given observations (x_i, y_i) and unknown parameters β_0,β_1 and σ^2 is

$$L(x_i, y_i; \beta_0, \beta_1, \sigma^2) = \prod_{i=1}^{n} \left(\frac{1}{2\pi\sigma^2} \right)^{1/2} \exp\left[-\frac{1}{2\sigma^2} (y_i - \beta_0 - \beta_1 x_i)^2 \right].$$

The maximum likelihood estimates of β_0, β_1 and σ^2 can be obtained by maximizing $L(x_i, y_i; \beta_0, \beta_1, \sigma^2)$ or equivalently $\ln L(x_i, y_i; \beta_0, \beta_1, \sigma^2)$ where

$$\ln L(x_i, y_i; \beta_0, \beta_1, \sigma^2) = -\left(\frac{n}{2}\right) \ln 2\pi - \left(\frac{n}{2}\right) \ln \sigma^2 - \left(\frac{1}{2\sigma^2}\right) \sum_{i=1}^{n} (y_i - \beta_0 - \beta_1 x_i)^2.$$

The normal equations are obtained by partial differentiation of log-likelihood with respect to β_0, β_1 and σ^2 equating them to zero

$$\frac{\partial \ln L(x_i, y_i; \beta_0, \beta_1, \sigma^2)}{\partial \beta_0} = -\frac{1}{\sigma^2} \sum_{i=1}^{n} (y_i - \beta_0 - \beta_1 x_i) = 0$$

$$\frac{\partial \ln L(x_i, y_i; \beta_0, \beta_1, \sigma^2)}{\partial \beta_1} = -\frac{1}{\sigma^2} \sum_{i=1}^{n} (y_i - \beta_0 - \beta_1 x_i) x_i = 0$$

and

$$\frac{\partial \ln L(x_i, y_i; \beta_0, \beta_1, \sigma^2)}{\partial \sigma^2} = -\frac{n}{2\sigma^2} + \frac{1}{2\sigma^4} \sum_{i=1}^{n} (y_i - \beta_0 - \beta_1 x_i)^2 = 0.$$

The solution of these normal equations give the maximum likelihood estimates of β_0, β_1 and σ^2 as

$$\tilde{b}_0 = \bar{y} - \tilde{b}_1 \bar{x}$$

$$\tilde{b}_1 = \frac{\sum_{i=1}^{n} (x_i - \bar{x})(y_i - \bar{y})}{\sum_{i=1}^{n} (x_i - \bar{x})^2} = \frac{S_{xy}}{S_{xx}}$$

and

$$\tilde{s}^2 = \frac{\sum_{i=1}^{n} (y_i - \tilde{b}_0 - \tilde{b}_1 x_i)^2}{n}$$

respectively.

It can be verified that the Hessian matrix of second order partial derivation of ln L with respect to β_0, β_1, and σ^2 is negative definite at and $\beta_0 = \tilde{b}_0, \beta_1 = \tilde{b}_1, \sigma^2 = \tilde{s}^2)$ which ensures that the likelihood function is maximized at these values.

Note that the least squares and maximum likelihood estimates of β_0 and β_1 are identical when disturbances are normally distributed. The least squares and maximum likelihood estimates of σ^2 are different. In fact, the least squares estimate of σ^2 is

$$s^2 = \frac{1}{n-2}\sum_{i=1}^{n}(y_i - \bar{y})^2$$

so that it is related to maximum likelihood estimate as $\tilde{s}^2 = \frac{1}{n-2}s^2$.

Thus $\tilde{b}_0$ and $\tilde{b}_1$ are unbiased estimators of β_0 and β_1 whereas $\tilde{s}^2$ is a biased estimate of σ^2, but it is asymptotically unbiased. The variances of $\tilde{b}_0$ and $\tilde{b}_1$ are same as that of b_0 and b_1 respectively but the mean squared error

$$MSE(\tilde{s}^2) < Var(s^2).$$

Testing of Hypotheses and Confidence Interval Estimation for Slope Parameter

Now we consider the tests of hypothesis and confidence interval estimation for the slope parameter of the model under two cases, viz., when σ^2 is known and when σ^2 is unknown.

Case 1 : When σ^2 is known

Consider the simple linear regression model $y_i = \beta_0 + \beta_1 x_i + \varepsilon_i$ $(i = 1, 2, ..., n).$ It is assumed that ε_i's are independent and identically distributed and follow $N(0, \sigma^2)$.

First we develop a test for the null hypothesis related to the slope parameter

$$H_0 : \beta_1 = \beta_{10}$$

where β_{10} is some given constant.

Assuming σ^2 to be known, we know that

$$E(b_1) = \beta_1, Var(b_1) = \frac{\sigma^2}{S_{xx}}$$

and b_1 is a linear combination of normally distributed y_i's, so

$$b_1 \sim N\left(\beta_1, \frac{\sigma^2}{S_{xx}}\right)$$

and so the following statistic can be constructed

$$Z_1 = \frac{b_1 - \beta_{10}}{\sqrt{\dfrac{\sigma^2}{S_{xx}}}}$$

which is distributed as N(0, 1) when H_o is true.

A decision rule to test $H_1 : \beta_1 \neq \beta_{10}$ can be framed as follows:

Reject H_o if $|Z_o| > z_{\alpha/2}$

where $z_{\alpha/2}$ is the $\alpha/2$ percentage points on normal distribution. Similarly, the decision rule for one sided alternative hypothesis can also be framed.

The 100(1   )% confidence interval for β_1 can be obtained using the Z_1 statistic as follows:

$$P\left[-z_{\frac{\alpha}{2}} \leq Z_1 \leq z_{\frac{\alpha}{2}}\right] = 1 - \alpha$$

$$P\left[-z_{\frac{\alpha}{2}} \leq \frac{b_1 - \beta_1}{\sqrt{\dfrac{\sigma^2}{S_{xx}}}} \leq z_{\frac{\alpha}{2}}\right] = 1 - \alpha$$

$$P\left[b_1 - z_{\frac{\alpha}{2}}\sqrt{\frac{\sigma^2}{S_{xx}}} \leq \beta_1 \leq b_1 + z_{\frac{\alpha}{2}}\sqrt{\frac{\sigma^2}{S_{xx}}}\right] = 1 - \alpha$$

So $100(1 - \alpha)\%$ confidence interval for β_1 is

$$\left(b_1 - z_{\alpha/2}\sqrt{\frac{\alpha^2}{S_{xx}}}, b_1 + z_{\alpha/2}\sqrt{\frac{\alpha^2}{S_{xx}}}\right)$$

where $z_{\alpha/2}$ is the $\alpha/2$ percentage point of the $N(0,1)$ distribution.

Case 2 : When σ^2 is unknown

When σ^2 is unknown, we proceed as follows. We know that

$$\frac{SS_{res}}{\sigma^2} \sim \chi^2(n-2).$$

And

$$E\left(\frac{SS_{res}}{n-2}\right) = \sigma^2.$$

Further, SS_{res}/σ^2 and b_1 are independently distributed. This result also follows from the result that under normal distribution, the maximum likelihood estimates, viz., sample mean (estimator of population mean) and sample variance (estimator of population variance) are independently distributed so b_1 and s^2 are also independently distributed.

Thus the following statistic can be constructed:

$$t_0 = \frac{b_1 - \beta_1}{\sqrt{\dfrac{\hat{\sigma}^2}{s_{xx}}}}$$

$$= \frac{b_1 - \beta_1}{\sqrt{\dfrac{SS_{res}}{(n-2)s_{xx}}}} \sim t_{n-2}$$

which follows a t-distribution with (n - 2) degrees of freedom, denoted as t_{n-2}, when H_0 is true.

A decision rule to test $H_1 : \beta_1 \neq \beta_{10}$ is to

reject H_0 if $|t_0| > t_{n-2,\alpha/2}$

where $t_{n-2,\alpha/2}$ is the $\alpha/2$ percentage point of the t-distribution with (n - 2) degrees of freedom.

Similarly, the decision rule for one sided alternative hypothesis can also be framed.

The $100(1-\alpha)\%$ confidence interval of β_1 can be obtained using the t_0 statistic as follows :

$$P\left[-t_{\frac{\alpha}{2}} \leq t_0 \leq t_{\frac{\alpha}{2}} \right] = 1 - \alpha$$

$$P\left[-t_{\frac{\alpha}{2}} \leq \frac{b_1 - \beta_1}{\sqrt{\dfrac{\hat{\sigma}^2}{s_{xx}}}} \leq t_{\frac{\alpha}{2}} \right] = 1 - \alpha$$

$$P\left[b_1 - t_{\frac{\alpha}{2}}\sqrt{\dfrac{\hat{\sigma}^2}{s_{xx}}} \leq \beta_1 \leq b_1 + t_{\frac{\alpha}{2}}\sqrt{\dfrac{\hat{\sigma}^2}{s_{xx}}} \right] = 1 - \alpha.$$

So $100(1-\alpha)\%$ the confidence interval β_1 is

$$\left(b_1 - t_{n-2,\alpha/2}\sqrt{\frac{SS_{res}}{(n-2)s_{xx}}}, b_1 + t_{n-2,\alpha/2}\sqrt{\frac{SS_{res}}{(n-2)s_{xx}}} \right).$$

Testing of Hypotheses and Confidence Interval Estimation for Intercept Term

Now, we consider the tests of hypothesis and confidence interval estimation for intercept term under two cases, viz., when σ^2 is known and when σ^2 is unknown.

Case 1: When σ^2 is known

Suppose the null hypothesis under consideration is $H_0 : \beta_0 = \beta_{00}$, where σ^2 is known, then using the result that $E(b_0) = \beta_0$, $Var(b_0) = \sigma^2 \left(\dfrac{1}{n} + \dfrac{\overline{x}^2}{S_x} \right)$ and b_0 is a linear combination of normally distributed random variables, the following statistic

$$Z_0 = \frac{b_0 - \beta_{00}}{\sqrt{\sigma^2 \left(\dfrac{1}{n} + \dfrac{\overline{x}^2}{S_{xx}} \right)}} \sim N(0,1),$$

has a N(0, 1) distribution when H_0 is true.

A decision rule to test $H_1 : \beta_0 \neq \beta_{00}$ can be framed as follows:

Reject H_0 if $|Z_0| > z_{\alpha/2}$

where $_{/2}$ is the $\alpha / 2$ percentage points on normal distribution.

Similarly, the decision rule for one sided alternative hypothesis can also be framed.

The $100(1-\alpha)\%$ confidence intervals for β_0 when σ^2 is known can be derived using the Z_0 statistic as follows:

$$P\left[-z_{\frac{\alpha}{2}} \leq Z_0 \leq z_{\frac{\alpha}{2}} \right] = 1 - \alpha$$

$$P\left[-z_{\frac{\alpha}{2}} \leq \frac{b_0 - \beta_0}{\sqrt{\left(\dfrac{1}{n} + \dfrac{\overline{x}^2}{S_{xx}} \right)}} \leq z_{\frac{\alpha}{2}} \right] = 1 - \alpha$$

$$P\left[b_0 - z_{\frac{\alpha}{2}} \sqrt{\sigma^2 \left(\dfrac{1}{n} + \dfrac{\overline{x}^2}{S_{xx}} \right)} \leq \beta_0 \leq b_0 + z_{\frac{\alpha}{2}} \sqrt{\sigma^2 \left(\dfrac{1}{n} + \dfrac{\overline{x}^2}{S_{xx}} \right)} \right] = 1 - \alpha.$$

So the $100(1-\alpha)\%$ of confidential interval of β_0 is

$$\left(b_0 - z_{\alpha/2} \sqrt{\sigma^2 \left(\dfrac{1}{n} + \dfrac{\overline{x}^2}{S_{xx}} \right)}, b_0 + z_{\alpha/2} \sqrt{\sigma^2 \left(\dfrac{1}{n} + \dfrac{\overline{x}^2}{S_{xx}} \right)} \right).$$

Case 2: When σ^2 is unknown

When σ^2 is unknown, then the statistic is constructed

$$t_0 = \frac{b_0 - \beta_{00}}{\sqrt{\dfrac{SS_{res}}{n-2}\left(\dfrac{1}{n} + \dfrac{\bar{x}^2}{S_{xx}}\right)}}$$

which follows a t-distribution with $(n - 2)$ degrees of freedom, i.e., t_{n-2} when H_o is true.

A decision rule to test $H_1 : \beta_0 \neq \beta_{00}$ is as follows:

Reject H_o whenever $|t_0| > t_{n-2,\alpha/2}$

Where $t_{n-2,\alpha/2}$ is the $\alpha/2$ percentage point of the t-distribution with $(n - 2)$ degrees of freedom.

Similarly, the decision rule for one sided alternative hypothesis can also be framed.

The $100(1-\alpha)\%$ of confidential interval of β_0 can be obtained as follows:

Consider

$$P\left[t_{n-2,\alpha/2} \leq t_0 \leq t_{n-2,\alpha/2}\right] = 1 - \alpha$$

$$P\left[t_{n-2,\alpha/2} \leq \frac{b_0 - \beta_0}{\sqrt{\dfrac{SS_{res}}{n-2}\left(\dfrac{1}{n} + \dfrac{\bar{x}^2}{S_{xx}}\right)}} \leq t_{n-2,\alpha/2}\right] = 1 - \alpha$$

$$P\left[b_0 - t_{n-2,\alpha/2}\sqrt{\dfrac{SS_{res}}{n-2}\left(\dfrac{1}{n} + \dfrac{\bar{x}^2}{S_{xx}}\right)} \leq \beta_0 \leq b_0 + t_{n-2,\alpha/2}\sqrt{\dfrac{SS_{res}}{n-2}\left(\dfrac{1}{n} + \dfrac{\bar{x}^2}{S_{xx}}\right)}\right] = 1 - \alpha.$$

The $100(1-\alpha)\%$ of confidential interval of β_0 is

$$\left[b_0 - t_{n-2,\alpha/2}\sqrt{\dfrac{SS_{res}}{n-2}\left(\dfrac{1}{n} + \dfrac{\bar{x}^2}{S_{xx}}\right)}, b_0 + t_{n-2,\alpha/2}\sqrt{\dfrac{SS_{res}}{n-2}\left(\dfrac{1}{n} + \dfrac{\bar{x}^2}{S_{xx}}\right)}\right].$$

Confidence interval for σ^2

A confidence interval for σ^2 can also be derived as follows. Since $SS_{res}/\sigma^2 \sim \chi^2_{n-2}$, thus consider

$$P\left[\chi^2_{n-2,\alpha/2} \leq \frac{SS_{res}}{\sigma^2} \leq \chi^2_{n-2,1-\alpha/2}\right] = 1 - \alpha$$

$$P\left[\frac{SS_{res}}{\chi^2_{n-2,1-\alpha/2}} \leq \sigma^2 \leq \frac{SS_{res}}{\chi^2_{n-2,\alpha/2}}\right] = 1 - \alpha.$$

The corresponding $100(1-\alpha)\%$ confidence interval for σ^2 is $\left(\dfrac{SS_{res}}{\chi^2_{n-2,1-\alpha/2}}, \dfrac{SS_{res}}{\chi^2_{n-2,\alpha/2}} \right)$.

Analysis of Variance

Biologist and statistician Ronald Fisher

Analysis of variance (ANOVA) is a collection of statistical models used to analyze the differences among group means and their associated procedures (such as "variation" among and between groups), developed by statistician and evolutionary biologist Ronald Fisher. In the ANOVA setting, the observed variance in a particular variable is partitioned into components attributable to different sources of variation. In its simplest form, ANOVA provides a statistical test of whether or not the means of several groups are equal, and therefore generalizes the t-test to more than two groups. ANOVAs are useful for comparing (testing) three or more means (groups or variables) for statistical significance. It is conceptually similar to multiple two-sample t-tests, but is more conservative (results in less type I error) and is therefore suited to a wide range of practical problems.

History

While the analysis of variance reached fruition in the 20th century, antecedents extend centuries into the past according to Stigler. These include hypothesis testing, the partitioning of sums of squares, experimental techniques and the additive model. Laplace was performing hypothesis testing in the 1770s. The development of least-squares methods by Laplace and Gauss circa 1800 provided an improved method of combining observations (over the existing practices then used in astronomy and geodesy). It also initiated much study of the contributions to sums of squares. Laplace soon knew how to estimate a variance from a residual (rather than a total) sum of squares. By 1827 Laplace was using least squares methods to address ANOVA problems regarding measurements of atmospheric tides. Before 1800 astronomers had isolated observational errors resulting from reaction times (the "personal equation") and had developed methods of reducing the errors. The experimental methods used in the study of the personal equation were later accepted by the emerging field of psychology which developed strong (full factorial) experimental methods to which randomization and blinding were soon added. An eloquent non-mathematical explanation of the additive effects model was available in 1885.

Ronald Fisher introduced the term variance and proposed its formal analysis in a 1918 article *The Correlation Between Relatives on the Supposition of Mendelian Inheritance.* His first application of the analysis of variance was published in 1921. Analysis of variance became widely known after being included in Fisher's 1925 book *Statistical Methods for Research Workers.*

Randomization models were developed by several researchers. The first was published in Polish by Neyman in 1923.

One of the attributes of ANOVA which ensured its early popularity was computational elegance. The structure of the additive model allows solution for the additive coefficients by simple algebra rather than by matrix calculations. In the era of mechanical calculators this simplicity was critical. The determination of statistical significance also required access to tables of the F function which were supplied by early statistics texts.

Motivating Example

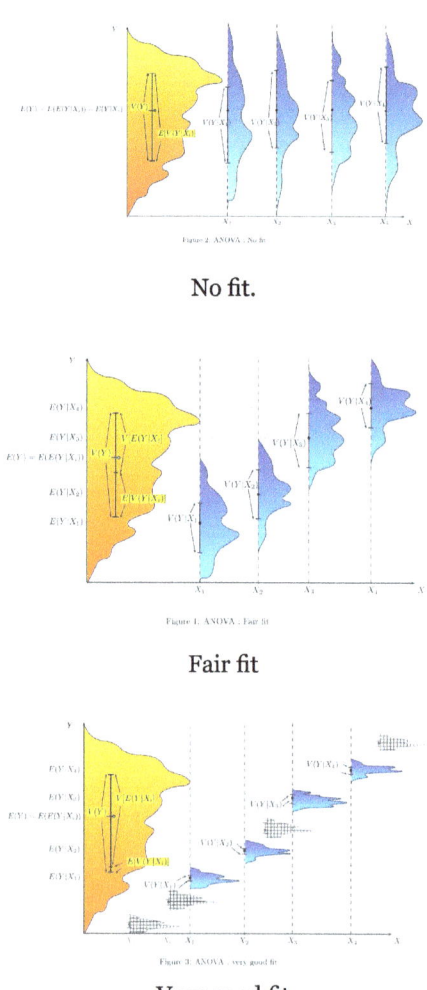

No fit.

Fair fit

Very good fit

The analysis of variance can be used as an exploratory tool to explain observations. A dog show provides an example. A dog show is not a random sampling of the breed: it is typically limited to dogs that are adult, pure-bred, and exemplary. A histogram of dog weights from a show might plausibly be rather complex, like the yellow-orange distribution shown in the illustrations. Suppose we wanted to predict the weight of a dog based on a certain set of characteristics of each dog. Before we could do that, we would need to *explain* the distribution of weights by dividing the dog population into groups based on those characteristics. A successful grouping will split dogs such that (a) each group has a low variance of dog weights (meaning the group is relatively homogeneous) and (b) the mean of each group is distinct (if two groups have the same mean, then it isn't reasonable to conclude that the groups are, in fact, separate in any meaningful way).

In the illustrations to the right, each group is identified as X_1, X_2, etc. In the first illustration, we divide the dogs according to the product (interaction) of two binary groupings: young vs old, and short-haired vs long-haired (thus, group 1 is young, short-haired dogs, group 2 is young, long-haired dogs, etc.). Since the distributions of dog weight within each of the groups (shown in blue) has a large variance, and since the means are very close across groups, grouping dogs by these characteristics does not produce an effective way to explain the variation in dog weights: knowing which group a dog is in does not allow us to make any reasonable statements as to what that dog's weight is likely to be. Thus, this grouping fails to *fit* the distribution we are trying to explain (yellow-orange).

An attempt to explain the weight distribution by grouping dogs as (pet vs working breed) and (less athletic vs more athletic) would probably be somewhat more successful (fair fit). The heaviest show dogs are likely to be big strong working breeds, while breeds kept as pets tend to be smaller and thus lighter. As shown by the second illustration, the distributions have variances that are considerably smaller than in the first case, and the means are more reasonably distinguishable. However, the significant overlap of distributions, for example, means that we cannot reliably say that X_1 and X_2 are truly distinct (i.e., it is perhaps reasonably likely that splitting dogs according to the flip of a coin—by pure chance—might produce distributions that look similar).

An attempt to explain weight by breed is likely to produce a very good fit. All Chihuahuas are light and all St Bernards are heavy. The difference in weights between Setters and Pointers does not justify separate breeds. The analysis of variance provides the formal tools to justify these intuitive judgments. A common use of the method is the analysis of experimental data or the development of models. The method has some advantages over correlation: not all of the data must be numeric and one result of the method is a judgment in the confidence in an explanatory relationship.

Background and Terminology

ANOVA is a particular form of statistical hypothesis testing heavily used in the analysis of experimental data. A test result (calculated from the null hypothesis and the sample) is called statistically significant if it is deemed unlikely to have occurred by chance, *assuming the truth of the null hypothesis*. A statistically significant result, when a probability (p-value) is less than a threshold (significance level), justifies the rejection of the null hypothesis, but only if the a priori probability of the null hypothesis is not high.

In the typical application of ANOVA, the null hypothesis is that all groups are simply random samples of the same population. For example, when studying the effect of different treatments on similar samples of patients, the null hypothesis would be that all treatments have the same effect (perhaps none). Rejecting the null hypothesis would imply that different treatments result in altered effects.

By construction, hypothesis testing limits the rate of Type I errors (false positives) to a significance level. Experimenters also wish to limit Type II errors (false negatives). The rate of Type II errors depends largely on sample size (the rate will increase for small numbers of samples), significance level (when the standard of proof is high, the chances of overlooking a discovery are also high) and effect size (a smaller effect size is more prone to Type II error).

The terminology of ANOVA is largely from the statistical design of experiments. The experimenter adjusts factors and measures responses in an attempt to determine an effect. Factors are assigned to experimental units by a combination of randomization and blocking to ensure the validity of the results. Blinding keeps the weighing impartial. Responses show a variability that is partially the result of the effect and is partially random error.

ANOVA is the synthesis of several ideas and it is used for multiple purposes. As a consequence, it is difficult to define concisely or precisely.

"Classical ANOVA for balanced data does three things at once:

1. As exploratory data analysis, an ANOVA is an organization of an additive data decomposition, and its sums of squares indicate the variance of each component of the decomposition (or, equivalently, each set of terms of a linear model).

2. Comparisons of mean squares, along with an F-test ... allow testing of a nested sequence of models.

3. Closely related to the ANOVA is a linear model fit with coefficient estimates and standard errors."

In short, ANOVA is a statistical tool used in several ways to develop and confirm an explanation for the observed data.

Additionally:

4. It is computationally elegant and relatively robust against violations of its assumptions.

5. ANOVA provides industrial strength (multiple sample comparison) statistical analysis.

6. It has been adapted to the analysis of a variety of experimental designs.

As a result: ANOVA "has long enjoyed the status of being the most used (some would say abused) statistical technique in psychological research." ANOVA "is probably the most useful technique in the field of statistical inference."

ANOVA is difficult to teach, particularly for complex experiments, with split-plot designs being notorious. In some cases the proper application of the method is best determined by problem pattern recognition followed by the consultation of a classic authoritative test.

Design-of-experiments Terms

Balanced design

An experimental design where all cells (i.e. treatment combinations) have the same number of observations.

Blocking

A schedule for conducting treatment combinations in an experimental study such that any effects on the experimental results due to a known change in raw materials, operators, machines, etc., become concentrated in the levels of the blocking variable. The reason for blocking is to isolate a systematic effect and prevent it from obscuring the main effects. Blocking is achieved by restricting randomization.

Design

A set of experimental runs which allows the fit of a particular model and the estimate of effects.

DOE

Design of experiments. An approach to problem solving involving collection of data that will support valid, defensible, and supportable conclusions.

Effect

How changing the settings of a factor changes the response. The effect of a single factor is also called a main effect.

Error

Unexplained variation in a collection of observations. DOE's typically require understanding of both random error and lack of fit error.

Experimental unit

The entity to which a specific treatment combination is applied.

Factors

Process inputs an investigator manipulates to cause a change in the output.

Lack-of-fit error

Error that occurs when the analysis omits one or more important terms or factors from the process model. Including replication in a DOE allows separation of experimental error into its components: lack of fit and random (pure) error.

Model

Mathematical relationship which relates changes in a given response to changes in one or more factors.

Random error

Error that occurs due to natural variation in the process. Random error is typically assumed to be normally distributed with zero mean and a constant variance. Random error is also called experimental error.

Randomization

A schedule for allocating treatment material and for conducting treatment combinations in a DOE such that the conditions in one run neither depend on the conditions of the previous run nor predict the conditions in the subsequent runs.

Replication

Performing the same treatment combination more than once. Including replication allows an estimate of the random error independent of any lack of fit error.

Responses

The output(s) of a process. Sometimes called dependent variable(s).

Treatment

A treatment is a specific combination of factor levels whose effect is to be compared with other treatments.

Classes of Models

There are three classes of models used in the analysis of variance, and these are outlined here.

Fixed-effects Models

The fixed-effects model (class I) of analysis of variance applies to situations in which the experimenter applies one or more treatments to the subjects of the experiment to see whether the response variable values change. This allows the experimenter to estimate the ranges of response variable values that the treatment would generate in the population as a whole.

Random-effects Models

Random effects model (class II) is used when the treatments are not fixed. This occurs when the various factor levels are sampled from a larger population. Because the levels themselves are random variables, some assumptions and the method of contrasting the treatments (a multi-variable generalization of simple differences) differ from the fixed-effects model.

Mixed-effects Models

A mixed-effects model (class III) contains experimental factors of both fixed and random-effects types, with appropriately different interpretations and analysis for the two types.

Example: Teaching experiments could be performed by a college or university department to find a good introductory textbook, with each text considered a treatment. The fixed-effects model would compare a list of candidate texts. The random-effects model would determine whether important differences exist among a list of randomly selected texts. The mixed-effects model would compare the (fixed) incumbent texts to randomly selected alternatives.

Defining fixed and random effects has proven elusive, with competing definitions arguably leading toward a linguistic quagmire.

Assumptions of ANOVA

The analysis of variance has been studied from several approaches, the most common of which uses a linear model that relates the response to the treatments and blocks. Note that the model is linear in parameters but may be nonlinear across factor levels. Interpretation is easy when data is balanced across factors but much deeper understanding is needed for unbalanced data.

Textbook Analysis Using a Normal Distribution

The analysis of variance can be presented in terms of a linear model, which makes the following assumptions about the probability distribution of the responses:

- Independence of observations – this is an assumption of the model that simplifies the statistical analysis.

- Normality – the distributions of the residuals are normal.

- Equality (or "homogeneity") of variances, called homoscedasticity — the variance of data in groups should be the same.

The separate assumptions of the textbook model imply that the errors are independently, identically, and normally distributed for fixed effects models, that is, that the errors (ε) are independent and

$$\varepsilon \sim N(0, \sigma^2).$$

Randomization-based Analysis

In a randomized controlled experiment, the treatments are randomly assigned to experimental units, following the experimental protocol. This randomization is objective and declared before the experiment is carried out. The objective random-assignment is used to test the significance of the null hypothesis, following the ideas of C. S. Peirce and Ronald Fisher. This design-based analysis was discussed and developed by Francis J. Anscombe at Rothamsted Experimental Station and by Oscar Kempthorne at Iowa State University. Kempthorne and his students make an assumption of *unit treatment additivity*, which is discussed in the books of Kempthorne and David R. Cox.

Unit-treatment Additivity

In its simplest form, the assumption of unit-treatment additivity states that the observed response $y_{i,j}$ from experimental unit i when receiving treatment j can be written as the sum of the unit's response y_i and the treatment-effect t_j, that is

$$y_{i,j} = y_i + t_j.$$

The assumption of unit-treatment additivity implies that, for every treatment j, the j th treatment has exactly the same effect t_j on every experiment unit.

The assumption of unit treatment additivity usually cannot be directly falsified, according to Cox and Kempthorne. However, many *consequences* of treatment-unit additivity can be falsified. For a randomized experiment, the assumption of unit-treatment additivity *implies* that the variance is constant for all treatments. Therefore, by contraposition, a necessary condition for unit-treatment additivity is that the variance is constant.

The use of unit treatment additivity and randomization is similar to the design-based inference that is standard in finite-population survey sampling.

Derived Linear Model

Kempthorne uses the randomization-distribution and the assumption of *unit treatment additivity* to produce a *derived linear model*, very similar to the textbook model discussed previously. The test statistics of this derived linear model are closely approximated by the test statistics of an appropriate normal linear model, according to approximation theorems and simulation studies. However, there are differences. For example, the randomization-based analysis results in a small but (strictly) negative correlation between the observations. In the randomization-based analysis, there is *no assumption* of a *normal* distribution and certainly *no assumption* of *independence*. On the contrary, *the observations are dependent*.

The randomization-based analysis has the disadvantage that its exposition involves tedious algebra and extensive time. Since the randomization-based analysis is complicated and is closely approximated by the approach using a normal linear model, most teachers emphasize the normal linear model approach. Few statisticians object to model-based analysis of balanced randomized experiments.

Statistical Models for Observational Data

However, when applied to data from non-randomized experiments or observational studies, model-based analysis lacks the warrant of randomization. For observational data, the derivation of confidence intervals must use *subjective* models, as emphasized by Ronald Fisher and his followers. In practice, the estimates of treatment-effects from observational studies generally are often inconsistent. In practice, "statistical models" and observational data are useful for suggesting hypotheses that should be treated very cautiously by the public.

Summary of Assumptions

The normal-model based ANOVA analysis assumes the independence, normality and homogeneity of the variances of the residuals. The randomization-based analysis assumes only the homogeneity of the variances of the residuals (as a consequence of unit-treatment additivity) and uses the randomization procedure of the experiment. Both these analyses require homoscedasticity, as an assumption for the normal-model analysis and as a consequence of randomization and additivity for the randomization-based analysis.

However, studies of processes that change variances rather than means (called dispersion effects) have been successfully conducted using ANOVA. There are *no* necessary assumptions for ANOVA in its full generality, but the F-test used for ANOVA hypothesis testing has assumptions and practical limitations which are of continuing interest.

Problems which do not satisfy the assumptions of ANOVA can often be transformed to satisfy the assumptions. The property of unit-treatment additivity is not invariant under a "change of scale", so statisticians often use transformations to achieve unit-treatment additivity. If the response variable is expected to follow a parametric family of probability distributions, then the statistician may specify (in the protocol for the experiment or observational study) that the responses be transformed to stabilize the variance. Also, a statistician may specify that logarithmic transforms be applied to the responses, which are believed to follow a multiplicative model. According to Cauchy's functional equation theorem, the logarithm is the only continuous transformation that transforms real multiplication to addition.

Characteristics of ANOVA

ANOVA is used in the analysis of comparative experiments, those in which only the difference in outcomes is of interest. The statistical significance of the experiment is determined by a ratio of two variances. This ratio is independent of several possible alterations to the experimental observations: Adding a constant to all observations does not alter significance. Multiplying all observations by a constant does not alter significance. So ANOVA statistical significance result is independent of constant bias and scaling errors as well as the units used in expressing observations. In the era of mechanical calculation it was common to subtract a constant from all observations (when equivalent to dropping leading digits) to simplify data entry. This is an example of data coding.

Logic of ANOVA

The calculations of ANOVA can be characterized as computing a number of means and variances, dividing two variances and comparing the ratio to a handbook value to determine statistical significance. Calculating a treatment effect is then trivial, "the effect of any treatment is estimated by taking the difference between the mean of the observations which receive the treatment and the general mean."

Partitioning of the Sum of Squares

ANOVA uses traditional standardized terminology. The definitional equation of sample variance is $s^2 = \dfrac{1}{n-1}\sum(y_i - \overline{y})^2$, where the divisor is called the degrees of freedom (DF), the summation is called the sum of squares (SS), the result is called the mean square (MS) and the squared terms are deviations from the sample mean. ANOVA estimates 3 sample variances: a total variance based on all the observation deviations from the grand mean, an error variance based on all the observation deviations from their appropriate treatment means, and a treatment variance. The treatment variance is based on the deviations of treatment means from the grand mean, the result being multiplied by the number of observations in each treatment to account for the difference between the variance of observations and the variance of means.

The fundamental technique is a partitioning of the total sum of squares SS into components related to the effects used in the model. For example, the model for a simplified ANOVA with one type of treatment at different levels.

$$SS_{Total} = SS_{Error} + SS_{Treatments}$$

The number of degrees of freedom DF can be partitioned in a similar way: one of these components (that for error) specifies a chi-squared distribution which describes the associated sum of squares, while the same is true for "treatments" if there is no treatment effect.

$$DF_{Total} = DF_{Error} + DF_{Treatments}$$

The F-test

The F-test is used for comparing the factors of the total deviation. For example, in one-way, or single-factor ANOVA, statistical significance is tested for by comparing the F test statistic

$$F = \frac{\text{variance between treatments}}{\text{variance within treatments}}$$

$$F = \frac{MS_{Treatments}}{MS_{Error}} = \frac{SS_{Treatments} / (I-1)}{SS_{Error} / (n_T - I)}$$

where MS is mean square, I = number of treatments and n_T = total number of cases to the F-distribution with $I-1$, $n_T - I$ degrees of freedom. Using the F-distribution is a natural candidate because the test statistic is the ratio of two scaled sums of squares each of which follows a scaled chi-squared distribution.

The expected value of F is $1 + n\sigma_{Treatment}^2 / \sigma_{Error}^2$ (where n is the treatment sample size) which is 1 for no treatment effect. As values of F increase above 1, the evidence is increasingly inconsistent with the null hypothesis. Two apparent experimental methods of increasing F are increasing the sample size and reducing the error variance by tight experimental controls.

There are two methods of concluding the ANOVA hypothesis test, both of which produce the same result:

- The textbook method is to compare the observed value of F with the critical value of F determined from tables. The critical value of F is a function of the degrees of freedom of the numerator and the denominator and the significance level (α). If $F \geq F_{Critical}$, the null hypothesis is rejected.

- The computer method calculates the probability (p-value) of a value of F greater than or equal to the observed value. The null hypothesis is rejected if this probability is less than or equal to the significance level (α).

The ANOVA F-test is known to be nearly optimal in the sense of minimizing false negative errors for a fixed rate of false positive errors (i.e. maximizing power for a fixed significance level). For example, to test the hypothesis that various medical treatments have exactly the same effect, the F-test's p-values closely approximate the permutation test's p-values: The approximation is particularly close when the design is balanced. Such permutation tests characterize tests with maximum power against all alternative hypotheses, as observed by Rosenbaum. The ANOVA F–test (of the null-hypothesis that all treatments have exactly the same effect) is recommended as a practical test, because of its robustness against many alternative distributions.

Extended Logic

ANOVA consists of separable parts; partitioning sources of variance and hypothesis testing can be used individually. ANOVA is used to support other statistical tools. Regression is first used to fit more complex models to data, then ANOVA is used to compare models with the objective of selecting simple(r) models that adequately describe the data. "Such models could be fit without any reference to ANOVA, but ANOVA tools could then be used to make some sense of the fitted models, and to test hypotheses about batches of coefficients." "[W]e think of the analysis of variance as a way of understanding and structuring multilevel models—not as an alternative to regression but as a tool for summarizing complex high-dimensional inferences ..."

ANOVA for a Single Factor

The simplest experiment suitable for ANOVA analysis is the completely randomized experiment with a single factor. More complex experiments with a single factor involve constraints on randomization and include completely randomized blocks and Latin squares (and variants: Graeco-Latin squares, etc.). The more complex experiments share many of the complexities of multiple factors. A relatively complete discussion of the analysis (models, data summaries, ANOVA table) of the completely randomized experiment is available.

ANOVA for Multiple Factors

ANOVA generalizes to the study of the effects of multiple factors. When the experiment includes observations at all combinations of levels of each factor, it is termed factorial. Factorial experiments are more efficient than a series of single factor experiments and the efficiency grows as the number of factors increases. Consequently, factorial designs are heavily used.

The use of ANOVA to study the effects of multiple factors has a complication. In a 3-way ANOVA with factors x, y and z, the ANOVA model includes terms for the main effects (x, y, z) and terms for interactions (xy, xz, yz, xyz). All terms require hypothesis tests. The proliferation of interaction terms increases the risk that some hypothesis test will produce a false positive by chance. Fortunately, experience says that high order interactions are rare. The ability to detect interactions is a major advantage of multiple factor ANOVA. Testing one factor at a time hides interactions, but produces apparently inconsistent experimental results.

Caution is advised when encountering interactions; Test interaction terms first and expand the analysis beyond ANOVA if interactions are found. Texts vary in their recommendations regarding the continuation of the ANOVA procedure after encountering an interaction. Interactions complicate the interpretation of experimental data. Neither the calculations of significance nor the estimated treatment effects can be taken at face value. "A significant interaction will often mask the significance of main effects." Graphical methods are recommended to enhance understanding. Regression is often useful. A lengthy discussion of interactions is available in Cox (1958). Some interactions can be removed (by transformations) while others cannot.

A variety of techniques are used with multiple factor ANOVA to reduce expense. One technique used in factorial designs is to minimize replication (possibly no replication with support of analytical trickery) and to combine groups when effects are found to be statistically (or practically) insignificant. An experiment with many insignificant factors may collapse into one with a few factors supported by many replications.

Worked Numeric Examples

Several fully worked numerical examples are available. A simple case uses one-way (a single factor) analysis. A more complex case uses two-way (two-factor) analysis.

Associated Analysis

Some analysis is required in support of the *design* of the experiment while other analysis is performed after changes in the factors are formally found to produce statistically significant changes in the responses. Because experimentation is iterative, the results of one experiment alter plans for following experiments.

Preparatory Analysis

The Number of Experimental Units

In the design of an experiment, the number of experimental units is planned to satisfy the goals of the experiment. Experimentation is often sequential.

Early experiments are often designed to provide mean-unbiased estimates of treatment effects and of experimental error. Later experiments are often designed to test a hypothesis that a treatment effect has an important magnitude; in this case, the number of experimental units is chosen so that the experiment is within budget and has adequate power, among other goals.

Reporting sample size analysis is generally required in psychology. "Provide information on sample size and the process that led to sample size decisions." The analysis, which is written in the experimental protocol before the experiment is conducted, is examined in grant applications and administrative review boards.

Besides the power analysis, there are less formal methods for selecting the number of experimental units. These include graphical methods based on limiting the probability of false negative errors, graphical methods based on an expected variation increase (above the residuals) and methods based on achieving a desired confident interval.

Power Analysis

Power analysis is often applied in the context of ANOVA in order to assess the probability of successfully rejecting the null hypothesis if we assume a certain ANOVA design, effect size in the population, sample size and significance level. Power analysis can assist in study design by determining what sample size would be required in order to have a reasonable chance of rejecting the null hypothesis when the alternative hypothesis is true.

Effect Size

Several standardized measures of effect have been proposed for ANOVA to summarize the strength of the association between a predictor(s) and the dependent variable (e.g., η^2, ω^2, or f^2) or the overall standardized difference (Ψ) of the complete model. Standardized effect-size estimates facilitate comparison of findings across studies and disciplines. However, while standardized effect sizes are commonly used in much of the professional literature, a non-standardized measure of effect size that has immediately "meaningful" units may be preferable for reporting purposes.

Follow-up Analysis

It is always appropriate to carefully consider outliers. They have a disproportionate impact on statistical conclusions and are often the result of errors.

Model Confirmation

It is prudent to verify that the assumptions of ANOVA have been met. Residuals are examined or analyzed to confirm homoscedasticity and gross normality. Residuals should have the appearance of (zero mean normal distribution) noise when plotted as a function of anything including time and modeled data values. Trends hint at interactions among factors or among observations. One rule of thumb: "If the largest standard deviation is less than twice the smallest standard deviation, we can use methods based on the assumption of equal standard deviations and our results will still be approximately correct."

Follow-up Tests

A statistically significant effect in ANOVA is often followed up with one or more different follow-up tests. This can be done in order to assess which groups are different from which other groups or to test various other focused hypotheses. Follow-up tests are often distinguished in terms of whether they are planned (a priori) or post hoc. Planned tests are determined before looking at the data and post hoc tests are performed after looking at the data.

Often one of the "treatments" is none, so the treatment group can act as a control. Dunnett's test (a modification of the t-test) tests whether each of the other treatment groups has the same mean as the control.

Post hoc tests such as Tukey's range test most commonly compare every group mean with every other group mean and typically incorporate some method of controlling for Type I errors. Comparisons, which are most commonly planned, can be either simple or compound. Simple comparisons compare one group mean with one other group mean. Compound comparisons typically compare two sets of groups means where one set has two or more groups (e.g., compare average group means of group A, B and C with group D). Comparisons can also look at tests of trend, such as linear and quadratic relationships, when the independent variable involves ordered levels.

Following ANOVA with pair-wise multiple-comparison tests has been criticized on several grounds. There are many such tests (10 in one table) and recommendations regarding their use are vague or conflicting.

Study Designs and ANOVAs

There are several types of ANOVA. Many statisticians base ANOVA on the design of the experiment, especially on the protocol that specifies the random assignment of treatments to subjects; the protocol's description of the assignment mechanism should include a specification of the structure of the treatments and of any blocking. It is also common to apply ANOVA to observational data using an appropriate statistical model.

Some popular designs use the following types of ANOVA:

- One-way ANOVA is used to test for differences among two or more independent groups (means),e.g. different levels of urea application in a crop, or different levels of antibiotic action on several bacterial species, or different levels of effect of some medicine on groups of patients. Typically, however, the one-way ANOVA is used to test for differences among at least three groups, since the two-group case can be covered by a t-test. When there are only two means to compare, the t-test and the ANOVA F-test are equivalent; the relation between ANOVA and t is given by $F = t^2$.

- Factorial ANOVA is used when the experimenter wants to study the interaction effects among the treatments.

- Repeated measures ANOVA is used when the same subjects are used for each treatment (e.g., in a longitudinal study).

- Multivariate analysis of variance (MANOVA) is used when there is more than one response variable.

ANOVA Cautions

Balanced experiments (those with an equal sample size for each treatment) are relatively easy to interpret; Unbalanced experiments offer more complexity. For single factor (one way) ANOVA, the adjustment for unbalanced data is easy, but the unbalanced analysis lacks both robustness and power. For more complex designs the lack of balance leads to further complications. "The orthogonality property of main effects and interactions present in balanced data does not carry over to the unbalanced case. This means that the usual analysis of variance techniques do not apply. Consequently, the analysis of unbalanced factorials is much more difficult than that for balanced designs." In the general case, "The analysis of variance can also be applied to unbalanced data, but then the sums of squares, mean squares, and F-ratios will depend on the order in which the sources of variation are considered." The simplest techniques for handling unbalanced data restore balance by either throwing out data or by synthesizing missing data. More complex techniques use regression.

ANOVA is (in part) a significance test. The American Psychological Association holds the view that simply reporting significance is insufficient and that reporting confidence bounds is preferred.

While ANOVA is conservative (in maintaining a significance level) against multiple comparisons in one dimension, it is not conservative against comparisons in multiple dimensions.

Generalizations

ANOVA is considered to be a special case of linear regression which in turn is a special case of the general linear model. All consider the observations to be the sum of a model (fit) and a residual (error) to be minimized.

The Kruskal–Wallis test and the Friedman test are nonparametric tests, which do not rely on an assumption of normality.

Connection to Linear Regression

Below we make clear the connection between multi-way ANOVA and linear regression. Linearly re-order the data so that k^{th} observation is associated with a response y_k and factors $Z_{k,b}$ where $b \in \{1, 2, \ldots, B\}$ denotes the different factors and B is the total number of factors. In one-way ANOVA $B = 1$ and in two-way ANOVA $B = 2$. Furthermore, we assume the bth factor has I_b levels. Now, we can one-hot encode the factors into the $\prod_{b=1}^{B} I_b$ dimensional vector v_k.

The one-hot encoding function $g_b : I_b \mapsto \{0,1\}^{I_b}$ is defined such that the ith entry of $g_b(Z_{k,b})$ is

$$g_b(Z_{k,b})_i = \begin{cases} 1 & \text{if } i = Z_{k,b} \\ 0 & \text{otherwise} \end{cases}$$

The vector v_k is the concatenation of all of the above vectors for all b. Thus, $v_k = [g_1(Z_{k,1}), g_2(Z_{k,2}), \ldots, g_B(Z_{k,B})]$. In order to obtain a fully general B-way interaction ANOVA we must also concatenate every additional interaction term in the vector v_k and then add an intercept term. Let that vector be x_k.

With this notation in place, we now have the exact connection with linear regression. We simply regress response y_k against the vector X_k. However, there is a concern about identifiability. In order to overcome such issues we assume that the sum of the parameters within each set of interactions is equal to zero. From here, one can use F-statistics or other methods to determine the relevance of the individual factors.

Example

We can consider the 2-way interaction example where we assume that the first factor has 2 levels and the second factor has 3 levels.

Define $a_i = 1$ if $Z_{k,1} = i$ and $b_i = 1$ if $Z_{k,2} = i$, i.e. a is the one-hot encoding of the first factor and b is the one-hot encoding of the second factor.

With that,

$$X_k = [a_1, a_2, b_1, b_2, b_3, a_1 \times b_1, a_1 \times b_2, a_1 \times b_3, a_2 \times b_1, a_2 \times b_2, a_2 \times b_3, 1]$$

where the last term is an intercept term. For a more concrete example suppose that

$$Z_{k,1} = 2$$
$$Z_{k,2} = 1$$

Then,

$$X_k = [0,1,1,0,0,0,0,0,1,0,0,1]$$

The technique of analysis of variance is usually used for testing the hypothesis related to equality of more than one parameters, like population means or slope parameters. It is more meaningful in case of multiple regression model when there are more than one slope parameters. This technique is discussed and illustrated here to understand the related basic concepts and fundamentals which will be used in developing the analysis of variance in multiple linear regression model where the explanatory variables are more than two.

A test statistic for testing $H_0 : \beta_1 = 0$ can also be formulated using the analysis of variance technique as follows.

On the basis of the identity $y_i - \hat{y}_i = (y_i - \bar{y}) - (\hat{y}_i - \bar{y})$,

the sum of squared residuals is

$$S(b) = \sum_{i=1}^{n} (y_i - \hat{y}_i)^2$$

$$= \sum_{i=1}^{n} (y_i - \bar{y})^2 + \sum_{i=1}^{n} (\hat{y}_i - \bar{y}_i)^2 - 2\sum_{i=1}^{n} (y_i - \bar{y})(\hat{y}_i - \bar{y}).$$

Further consider

$$\sum_{i=1}^{n} (y_i - \bar{y})(\hat{y}_i - \bar{y}) = \sum_{i=1}^{n} (y_i - \bar{y})b_1(x_i - \bar{x})$$

$$= b_1^2 \sum_{i=1}^{n} (x_i - \bar{x})^2$$

$$= \sum_{i=1}^{n} (\hat{y}_i - \bar{y})^2.$$

Thus we have $\sum_{i=1}^{n} (y_i - \bar{y})^2 = \sum_{i=1}^{n} (y_i - \hat{y}_i)^2 + \sum_{i=1}^{n} (\hat{y}_i - \bar{y})^2.$

The term $\sum_{i=1}^{n} (y_i - \bar{y})^2$ is called the sum of squares about the mean or corrected sum of squares of y (i.e., SS corrected) or total sum of squares denoted as s_{yy}.

The term $\sum_{i=1}^{n} (y_i - \hat{y}_i)^2$ describes the deviation: observation minus predicted value, viz., the residual sum of squares, i.e.

$$SS_{res} = \sum_{i=1}^{n} (y_i - \hat{y}_i)^2$$

whereas the term $\sum_{i=1}^{n} (\hat{y}_i - \overline{y})^2$ describes the proportion of variability explained by regression,

$$SS_{reg} = \sum_{i=1}^{n} (\hat{y}_i - \overline{y})^2.$$

If all observations y_i are located on a straight line, then in this case $\sum_{i=1}^{n}(y_i - \hat{y}_i)^2 = 0$ and thus $SS_{corrected} = SS_{res}$.

Note that SS_{reg} is completely determined by b_1 and so has only one degrees of freedom. The total sum of squares $s_{yy} = \sum_{i=1}^{n}(y_i - \overline{y})^2$ has (n - 1) degrees of freedom due to constraint $\sum_{i=1}^{n}(y_i - \overline{y}) = 0$ and SS_{res} has (n - 2) degrees of freedom as it depends on b_0 and b_1.

All sums of squares are mutually independent and distributed as χ_{df}^2 with df degrees of freedom if the errors are normally distributed.

The mean square due to regression is

$$MS_{reg} = \frac{SS_{res}}{1}$$

and mean square due to residuals is

$$MSE = \frac{SS_{res}}{n-2}.$$

The test statistic for testing $H_0 : \beta_1 = 0$ is

$$F_0 = \frac{MS_{reg}}{MSE}.$$

If $H_0 : \beta_1 = 0$ is true, then MS_{reg} and MSE are independently distributed and thus $F_0 \sim F_{1,n-2}$.

The decision rule for $H_1 : \beta_1 \neq 0$ is to reject H_0 if $F_0 > F_{1,n-2;1-\alpha}$ at α level of significance. The test procedure can be described in an Analysis of Variance table.

Analysis of variance for testing $H_0 : \beta_1 = 0$			
Source of variation	Sum of squares	Degrees of freedom	Mean Square
Regression	SS_{reg}	1	MS_{reg}

Residual	SS_{res}	n − 2	MSE
Total	s_{yy}	n-1	

Some other forms of SS_{reg}, SS_{res} and s_{yy} can be derived as follows:

The sample correlation coefficient then may be written as

$$xy \qquad \frac{\overline{xy}}{\sqrt{S_{xx}}\sqrt{S_{yy}}}$$

Moreover, we have

$$b_1 = \frac{S_{xy}}{S_{xx}} = r_{xy}\sqrt{\frac{S_{yy}}{S_{xx}}}.$$

The estimator of σ^2 in this case may be expressed as

$$s^2 = \frac{1}{n-2}\sum_{i=1}^{n} e_i^2$$

$$= \frac{1}{n-2} SS_{res}.$$

Various alternative formulations for SS_{res} are in use as well:

$$SS_{res} = \sum_{i=1}^{n}\left[y_i - (b_0 + b_1 x_i)\right]^2$$

$$= \sum_{i=1}^{n}\left[(y_i - \overline{y}) - b_1(x_i - \overline{x})\right]^2$$

$$= s_{yy} + b_1^2 s_{xx} - 2b_1 s_{xy}$$

$$= s_{yy} - b_1^2 s_{xx}$$

$$= s_{yy} - \frac{(s_{xy})^2}{s_{xx}}.$$

Using this result, we find that

$$SS_{corrected} = s_{yy}$$

and

$$SS_{reg} = s_{yy} - SS_{res}$$

$$= \frac{(s_{xy})^2}{s_{xx}}$$

$$= b_1^2 s_{xx}$$

$$= b_1 s_{xy}.$$

Goodness of Fit of Regression

It can be noted that a fitted model can be said to be good when residuals are small. Since SS_{res} is based on residuals, so a measure of quality of fitted model can be based on Ss_{res}. When intercept term is present in the model, a measure of goodness of fit of the model is given by

$$R^2 = 1 - \frac{SS_{res}}{s_{yy}}$$

$$= \frac{SS_{reg}}{s_{yy}}.$$

This is known as the coefficient of determination. This measure is based on the concept that how much variation in y's stated by s_{yy} is explainable by SS_{reg} and how much unexplainable part is contained in SS_{res}. The ratio SS_{reg} / s_{yy} describes the proportion of variability that is explained by regression in relation to the total variability of y. The ratio SS_{res} / s_{yy} describes the proportion of variability that is not covered by the regression.

It can be seen that

$$R^2 = r_{xy}^2.$$

where r_{xy} is the simple correlation coefficient between x and y. Clearly $0 \le R^2 \le 1$, so a value of R^2 closer to one indicates the better fit and value of R2 closer to zero indicates the poor fit.

Prediction Interval

In statistical inference, specifically predictive inference, a prediction interval is an estimate of an interval in which future observations will fall, with a certain probability, given what has already been observed. Prediction intervals are often used in regression analysis.

Prediction intervals are used in both frequentist statistics and Bayesian statistics: a prediction interval bears the same relationship to a future observation that a frequentist confidence interval or Bayesian credible interval bears to an unobservable population parameter: prediction intervals predict the distribution of individual future points, whereas confidence intervals and credible intervals of parameters predict the distribution of estimates of the true population mean or other quantity of interest that cannot be observed.

For example, if one makes the parametric assumption that the underlying distribution is a normal distribution, and has a sample set $\{X_1, ..., X_n\}$, then confidence intervals and credible intervals may be used to estimate the population mean μ and population standard deviation σ of the underlying population, while prediction intervals may be used to estimate the value of the next sample variable, X_{n+1}.

Alternatively, in Bayesian terms, a prediction interval can be described as a credible interval for the variable itself, rather than for a parameter of the distribution thereof.

The concept of prediction intervals need not be restricted to inference about a single future sample value but can be extended to more complicated cases. For example, in the context of river flooding where analyses are often based on annual values of the largest flow within the year, there may be interest in making inferences about the largest flood likely to be experienced within the next 50 years.

Since prediction intervals are only concerned with past and future observations, rather than unobservable population parameters, they are advocated as a better method than confidence intervals by some statisticians, such as Seymour Geisser, following the focus on observables by Bruno de Finetti.

Normal Distribution

Given a sample from a normal distribution, whose parameters are unknown, it is possible to give prediction intervals in the frequentist sense, i.e., an interval $[a, b]$ based on statistics of the sample such that on repeated experiments, X_{n+1} falls in the interval the desired percentage of the time; one may call these "predictive confidence intervals".

A general technique of frequentist prediction intervals is to find and compute a pivotal quantity of the observables X_1, ..., X_n, X_{n+1} – meaning a function of observables and parameters whose probability distribution does not depend on the parameters – that can be inverted to give a probability of the future observation X_{n+1} falling in some interval computed in terms of the observed values so far, $X_1, ..., X_n$. Such a pivotal quantity, depending only on observables, is called an ancillary statistic. The usual method of constructing pivotal quantities is to take the difference of two variables that depend on location, so that location cancels out, and then take the ratio of two variables that depend on scale, so that scale cancels out. The most familiar pivotal quantity is the Student's t-statistic, which can be derived by this method and is used in the sequel.

Known Mean, Known Variance

A prediction interval $[\ell, u]$ for a future observation X in a normal distribution $N(\mu, \sigma^2)$ with known mean and variance may easily be calculated from

$$\gamma = P(\ell < X < u) = P\left(\frac{\ell-\mu}{\sigma} < \frac{X-\mu}{\sigma} < \frac{u-\mu}{\sigma}\right) = P\left(\frac{\ell-\mu}{\sigma} < Z < \frac{u-\mu}{\sigma}\right),$$

where $Z = \dfrac{X-\mu}{\sigma}$, the standard score of X, is standard normal distributed.

Hence

$$\frac{\ell-\mu}{\sigma} = -z, \frac{u-\mu}{\sigma} = z,$$

or

$$\ell = \mu - z\sigma, u = \mu + z\sigma,$$

with z the quantile in the standard normal distribution for which:

$$\gamma = P(-z < Z < z).$$

or equivalently;

$$\tfrac{1}{2}(1-\gamma) = P(Z > z).$$

Prediction interval	z
75%	1.15
90%	1.64
95%	1.96
99%	2.58

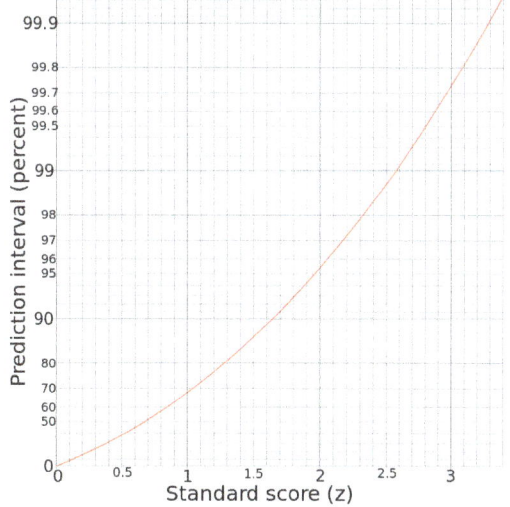

Prediction interval (on the y-axis) given from z (the quantile of the standard score, on the x-axis).
The y-axis is logarithmically compressed (but the values on it are not modified).

The prediction interval is conventionally written as:

$$\left[\mu - z\sigma, \mu + z\sigma\right].$$

For example, to calculate the 95% prediction interval for a normal distribution with a mean (μ) of 5 and a standard deviation (σ) of 1, then z is approximately 2. Therefore, the lower limit of the prediction interval is approximately $5 - (2 \cdot 1) = 3$, and the upper limit is approximately $5 + (2 \cdot 1) = 7$, thus giving a prediction interval of approximately 3 to 7.

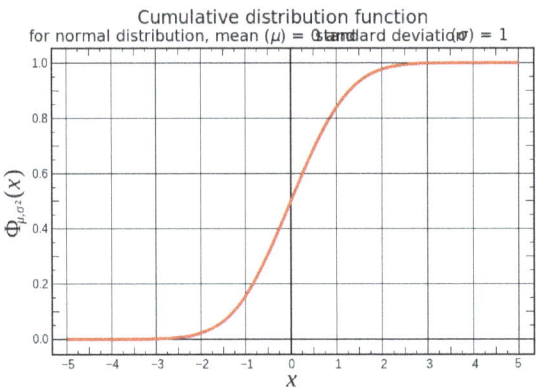

Diagram showing the cumulative distribution function for the normal distribution with mean (μ) 0 and variance (σ^2) 1. In addition to the quantile function, the prediction interval for any standard score can be calculated by $(1 - (1 - \Phi_{\mu,\sigma}{}^2(\text{standard score})) \cdot 2)$. For example, a standard score of $x = 1.96$ gives $\Phi_{\mu,\sigma}{}^2(1.96) = 0.9750$ corresponding to a prediction interval of $(1 - (1 - 0.9750) \cdot 2) = 0.9500 = 95\%$.

Estimation of Parameters

For a distribution with unknown parameters, a direct approach to prediction is to estimate the parameters and then use the associated quantile function – for example, one could use the sample mean $\bar{X}$ as estimate for μ and the sample variance s^2 as an estimate for σ^2. Note that there are two natural choices for s^2 here – dividing by $(n-1)$ yields an unbiased estimate, while dividing by n yields the maximum likelihood estimator, and either might be used. One then uses the quantile function with these estimated parameters $\Phi^{-1}_{\bar{X},s^2}$ to give a prediction interval.

This approach is usable, but the resulting interval will not have the repeated sampling interpretation – it is not a predictive confidence interval.

For the sequel, use the sample mean:

$$\bar{X} = \bar{X}_n = (X_1 + \cdots + X_n)/n$$

and the (unbiased) sample variance:

$$s^2 = s_n^2 = \frac{1}{n-1}\sum_{i=1}^{n}(X_i - \bar{X}_n)^2.$$

Unknown Mean, Known Variance

Given a normal distribution with unknown mean μ but known variance 1, the sample mean $\bar{X}$ of the observations $X_1, \ldots, X_n$ has distribution $N(\mu, 1/n)$, while the future observation X_{n+1} has distribution $N(\mu, 1)$. Taking the difference of these cancels the μ and yields a normal distribution of variance $1 + (1/n)$, thus

$$\frac{X_{n+1} - \bar{X}}{\sqrt{1 + (1/n)}} \sim N(0,1).$$

Solving for X_{n+1} gives the prediction distribution $N(\bar{X}, 1 + (1/n))$, from which one can compute intervals as before. This is a predictive confidence interval in the sense that if one uses a quantile range of $100p\%$, then on repeated applications of this computation, the future observation X_{n+1} will fall in the predicted interval $100p\%$ of the time.

Notice that this prediction distribution is more conservative than using the estimated mean $\bar{X}$ and known variance 1, as this uses variance $1 + (1/n)$, hence yields wider intervals. This is necessary for the desired confidence interval property to hold.

Known Mean, Unknown Variance

Conversely, given a normal distribution with known mean 0 but unknown variance σ^2, the sample variance s^2 of the observations $X_1, \ldots, X_n$ has, up to scale, a χ^2_{n-1} distribution; more precisely:

$$\frac{(n-1)s_n^2}{\sigma^2} \sim \chi^2_{n-1}.$$

while the future observation X_{n+1} has distribution $N(0, \sigma^2)$. Taking the ratio of the future observation and the sample standard deviation cancels the σ, yielding a Student's t-distribution with $n-1$ degrees of freedom:

$$\frac{X_{n+1}}{s} \sim T^{n-1}.$$

Solving for X_{n+1} gives the prediction distribution sT^{n-1}, from which one can compute intervals as before.

Notice that this prediction distribution is more conservative than using a normal distribution with the estimated standard deviation s and known mean 0, as it uses the t-distribution instead of the normal distribution, hence yields wider intervals. This is necessary for the desired confidence interval property to hold.

Unknown Mean, Unknown Variance

Combining the above for a normal distribution $N(\mu, \sigma^2)$ with both μ and σ^2 unknown yields the following ancillary statistic:

$$\frac{X_{n+1} - \bar{X}_n}{s_n \sqrt{1 + 1/n}} \sim T^{n-1}.$$

This simple combination is possible because the sample mean and sample variance of the normal distribution are independent statistics; this is only true for the normal distribution, and in fact characterizes the normal distribution.

Solving for X_{n+1} yields the prediction distribution

$$\bar{X}_n + s_n \sqrt{1 + 1/n} \cdot T^{n-1}.$$

The probability of X_{n+1} falling in a given interval is then:

$$\Pr\left(\bar{X}_n - T_a s_n \sqrt{1 + (1/n)} \leq X_{n+1} \leq \bar{X}_n + T_a s_n \sqrt{1 + (1/n)} \right) = p$$

where T_a is the $100(1 - p/2)^{\text{th}}$ percentile of Student's t-distribution with $n - 1$ degrees of freedom. Therefore the numbers

$$\bar{X}_n \pm T_a s_n \sqrt{1 + (1/n)}$$

are the endpoints of a $100(1 - p)\%$ prediction interval for X_{n+1}.

Non-parametric Methods

One can compute prediction intervals without any assumptions on the population; formally, this is a non-parametric method.

Suppose one randomly draws a sample of two observations X_1 and X_2 from a population in which values are assumed to have a continuous probability distribution

What is the probability that $X_2 > X_1$?

The answer is exactly 50%, *regardless* of the underlying population – the probability of picking 3 and then 7 is the same as picking 7 and then 3, regardless of the particular probability of picking 3 or 7. Thus, if one picks a single sample X_1, then 50% of the time the next sample will be greater, which yields $(X_1, +\infty)$ as a 50% prediction interval for X_2. Similarly, 50% of the time it will be smaller, which yields another 50% prediction interval for X_2, namely $(-\infty, X_1)$. Note that the assumption of a continuous distribution avoids the possibility that values might be exactly equal; this would complicate matters.

Similarly, if one has a sample $\{X_1, \ldots, X_n\}$ then the probability that the next observation X_{n+1} will be the largest is $1/(n + 1)$, since all observations have equal probability of being the maximum. In the same way, the probability that X_{n+1} will be the smallest is $1/(n + 1)$. The other $(n - 1)/(n + 1)$ of the time, X_{n+1} falls between the sample maximum and sample minimum of the sample $\{X_1, \ldots, X_n\}$. Thus, denoting the sample maximum and minimum by M and m, this yields an $(n - 1)/(n + 1)$ prediction interval of $[m, M]$.

For example, if $n = 19$, then $[m, M]$ gives an $18/20 = 90\%$ prediction interval – 90% of the time, the 20th observation falls between the smallest and largest observation seen heretofore. Likewise, $n = 39$ gives a 95% prediction interval, and $n = 199$ gives a 99% prediction interval.

More generally, if $X_{(j)}$ and $X_{(k)}$ are order statistics of the sample with $j < k$ and $j + k = n + 1$, then $[X_{(j)}, X_{(k)}]$ is a prediction interval for X_{n+1} with coverage probability (significance level) equal to $(n + 1 - 2j) / (n + 1)$.

One can visualize this by drawing the n samples on a line, which divides the line into $n + 1$ sections

(n − 1 segments between samples, and 2 intervals going to infinity at both ends), and noting that X_{n+1} has an equal chance of landing in any of these $n + 1$ sections. Thus one can also pick any k of these sections and give a $k/(n + 1)$ prediction interval (or set, if the sections are not consecutive). For instance, if $n = 2$, then the probability that X_3 will land between the existing two observations is 1/3.

Notice that while this gives the probability that a future observation will fall in a range, it does not give any estimate as to where in a segment it will fall – notably, if it falls outside the range of observed values, it may be far outside the range. Formally, this applies not just to sampling from a population, but to any exchangeable sequence of random variables, not necessarily independent or identically distributed.

Contrast with Confidence Intervals

Note that in the formula for the predictive confidence interval *no mention* is made of the unobservable parameters μ and σ of population mean and standard deviation – the observed *sample* statistics $\bar{X}_n$ and S_n of sample mean and standard deviation are used, and what is estimated is the outcome of *future* samples.

Rather than using sample statistics as estimators of population parameters and applying confidence intervals to these estimates, one considers "the next sample" X_{n+1} as *itself* a statistic, and computes its sampling distribution.

In parameter confidence intervals, one estimates population parameters; if one wishes to interpret this as prediction of the next sample, one models "the next sample" as a draw from this estimated population, using the (estimated) *population* distribution. By contrast, in predictive confidence intervals, one uses the *sampling* distribution of (a statistic of) a sample of n or $n + 1$ observations from such a population, and the population distribution is not directly used, though the assumption about its form (though not the values of its parameters) is used in computing the sampling distribution.

Applications

Prediction intervals are commonly used as definitions of reference ranges, such as reference ranges for blood tests to give an idea of whether a blood test is normal or not. For this purpose, the most commonly used prediction interval is the 95% prediction interval, and a reference range based on it can be called a *standard reference range*.

Regression Analysis

A common application of prediction intervals is to regression analysis.

Suppose the data is being modeled by a straight line regression:

$$y_i = \alpha + \beta x_i + \varepsilon_i$$

where y_i is the response variable, x_i is the explanatory variable, ε_i is a random error term, and α and β are parameters.

Given estimates $\hat{\alpha}$ and $\hat{\beta}$ for the parameters, such as from a simple linear regression, the predicted response value y_d for a given explanatory value x_d is

$$\hat{y}_d = \hat{\alpha} + \hat{\beta} x_d,$$

(the point on the regression line), while the actual response would be

$$y_d = \alpha + \beta x_d + \varepsilon_d.$$

The point estimate $\hat{y}_d$ is called the mean response, and is an estimate of the expected value of y_d, $E(y \mid x_d)$.

A prediction interval instead gives an interval in which one expects y_d to fall; this is not necessary if the actual parameters α and β are known (together with the error term ε_i), but if one is estimating from a sample, then one may use the standard error of the estimates for the intercept and slope ($\hat{\alpha}$ and $\hat{\beta}$), as well as their correlation, to compute a prediction interval.

In regression, Faraway (2002) makes a distinction between intervals for predictions of the mean response vs. for predictions of observed response—affecting essentially the inclusion or not of the unity term within the square root in the expansion factors above.

Bayesian Statistics

Seymour Geisser, a proponent of predictive inference, gives predictive applications of Bayesian statistics.

In Bayesian statistics, one can compute (Bayesian) prediction intervals from the posterior probability of the random variable, as a credible interval. In theoretical work, credible intervals are not often calculated for the prediction of future events, but for inference of parameters – i.e., credible intervals of a parameter, not for the outcomes of the variable itself. However, particularly where applications are concerned with possible extreme values of yet to be observed cases, credible intervals for such values can be of practical importance.

Prediction Interval Estimation

The $100(1-\alpha)\%$ prediction interval for $E(y \mid x_0)$ is obtained as follows:

The predictor $\hat{\mu}_{y|x_0}$ is a linear combination of normally distributed random variables, so it is also normally distributed as

$$\hat{\mu}_{y|x_0} \sim N(\beta_0 + \beta_1 x_0, PV(\hat{\mu}_{y|x_0})).$$

So if σ^2 is known, then the distribution of

$$\frac{\hat{\mu}_{y|x_0} - E(y \mid x_0)}{\sqrt{PV(\hat{\mu}_{y|x_0})}}$$

is N(0,1), so the $100(1-\alpha)\%$ prediction interval is obtained as

$$P\left[-z_{\frac{\alpha}{2}} \le \frac{\hat{\mu}_{y|x_0} - E(y|x_0)}{\sqrt{PV(\hat{\mu}_{y|x_0})}} \le z_{\frac{\alpha}{2}}\right] = 1-\alpha$$

which gives the prediction interval for $E(y|x_0)$ as

$$\left[\hat{\mu}_{y|x_0} - z_{\frac{\alpha}{2}}\sqrt{\sigma^2\left[\frac{1}{n} + \frac{(x_0 - \bar{x})^2}{S_{xx}}\right]}, \hat{\mu}_{y|x_0} + z_{\frac{\alpha}{2}}\sigma^2\left[\frac{1}{n} + \frac{(x_0 - \bar{x})^2}{S_{xx}}\right]\right].$$

When σ^2 is unknown, it is replaced by $\hat{\sigma}^2 = MSE$ and in this case the sampling distribution of

$$\frac{\hat{\mu}_{y|x_0} - E(y|x_0)}{\sqrt{MSE\left[\frac{1}{n} + \frac{(x_0 - \bar{x})^2}{S_{xx}}\right]}}$$

is t-distribution with (n - 2) degrees of freedom, i.e., t_{n-2}.

The $100(1-\alpha)\%$ prediction interval in this case is

$$P\left[-t_{\frac{\alpha}{2},n-2} \le \frac{\hat{\mu}_{y|x_0} - E(y|x_0)}{\sqrt{MSE\left[\frac{1}{n} + \frac{(x_0 - \bar{x})^2}{S_{xx}}\right]}} \le t_{\frac{\alpha}{2},n-2}\right] = 1-\alpha$$

which gives the prediction interval as

$$\left(\hat{\mu}_{y|x_0} - t_{\frac{\alpha}{2},n-2}\sqrt{MSE\left(\frac{1}{n} + \frac{(x_0 - \bar{x})^2}{S_{xx}}\right)}, \hat{\mu}_{y|x_0} + t_{\frac{\alpha}{2},n-2}\sqrt{MSE\left(\frac{1}{n} + \frac{(x_0 - \bar{x})^2}{S_{xx}}\right)}\right).$$

Note that the width of prediction interval $E(y|x_0)$ is a function of x_0. The interval width is minimum for $x_0 = \bar{x}$ and widens as $|x_0 - \bar{x}|$ increases. This is expected also as the best estimates of y to be made at x-values lie near the center of the data and the precision of estimation to deteriorate as we move to the boundary of the x-space.

Prediction of Actual Value

If x_0 is the value of the explanatory variable, then the actual value predictor for y is .

$$\hat{y}_0 = b_0 + b_1 x_0.$$

Note that the form of predictor is same as of average value predictor but its predictive error and other properties are different. This is the dual nature of predictor.

Predictive Bias

Then the prediction error of $\hat{y}_0$ is given as

$$\hat{y}_0 - y_0 = b_0 + b_1 x_0 - (\beta_0 + \beta_1 x_0 + \varepsilon)$$
$$= (b_0 - \beta_0) + (b_1 - \beta_1) x_0 - \varepsilon.$$

Thus, we find that

$$E(\hat{y}_0 - y_0) = E(b_0 - \beta_0) + E(b_1 - \beta_1) x_0 - E(\varepsilon)$$
$$= 0 + 0 + 0 = 0$$

which implies that $\hat{y}_0$ is an unbiased predictor of y.

Predictive Variance

Because the future observation y_0 is independent of $\hat{y}_0$, the predictive variance of $\hat{y}_0$ is

$$PV(\hat{y}_0) = E(\hat{y}_0 - y_0)^2$$
$$= E[(b_0 - \beta_0) + (x_0 - \bar{x})(b_1 - \beta_1) + (b_1 - \beta_1)\bar{x} - \varepsilon_0]^2$$
$$= Var(b_0) + (x_0 - \bar{x})^2 Var(b_1) + \bar{x}^2 Var(b_1) + Var(\varepsilon) + 2(x_0 - \bar{x})Cov(b_0, b_1) + 2\bar{x}Cov(b_0, b_1) + 2(x_0 - \bar{x})Var(b_1)$$

[rest of the terms are 0 assuming the independence of ε_0 with $\varepsilon_1, \varepsilon_2, ..., \varepsilon_n$]

$$= Var(b_0) + [(x_0 - \bar{x})^2 + \bar{x}^2 + 2(x_0 - \bar{x})]Var(b_1) + Var(\varepsilon) + 2[(x_0 - \bar{x}) + 2\bar{x}]Cov(b_0, b_1)$$
$$= Var(b_0) + x_0^2 Var(b_1) + Var(\varepsilon) + 2x_0 Cov(b_0, b_1)$$
$$= \sigma^2 \left[\frac{1}{n} + \frac{\bar{x}^2}{S_{xx}} \right] + x_0^2 \frac{\sigma^2}{S_{xx}} + \sigma^2 - 2x_0 \frac{\bar{x}\sigma^2}{S_{xx}}$$
$$= \sigma^2 \left[1 + \frac{1}{n} + \frac{(x_0 - \bar{x})^2}{S_{xx}} \right].$$

Estimate of Predictive Variance

The estimate of predictive variance can be obtained by replacing σ^2 by its estimate $\hat{\sigma}^2 = MSE$ as

$$\widehat{PV}(\hat{y}_0) = \hat{\sigma}^2 \left[1 + \frac{1}{n} + \frac{(x_0 - \bar{x})^2}{S_{xx}} \right]$$
$$= MSE \left[1 + \frac{1}{n} + \frac{(x_0 - \bar{x})^2}{S_{xx}} \right].$$

Prediction Interval

If σ^2 is known, then the distribution of

$$\frac{\hat{y}_0 - E(\hat{y}_0)}{\sqrt{PV(\hat{y}_0)}}$$

is $N(0, 1)$. So the $100(1-\alpha)\%$ prediction interval is obtained as

$$P\left[-z_{\frac{\alpha}{2}} \leq \frac{\hat{y}_0 - E(\hat{y}_0)}{\sqrt{PV(\hat{y}_0)}} \leq z_{\frac{\alpha}{2}}\right] = 1-\alpha$$

which gives the prediction interval for $\hat{y}_0$ as

$$\left(\hat{y}_0 - z_{\frac{\alpha}{2}}\sqrt{\sigma^2\left(1+\frac{1}{n}+\frac{(x_0-\overline{x})^2}{S_{xx}}\right)},\ \hat{y}_0 + z_{\frac{\alpha}{2}}\sqrt{\sigma^2\left(1+\frac{1}{n}+\frac{(x_0-\overline{x})^2}{S_{xx}}\right)}\right).$$

When σ^2 is unknown, then

$$\frac{\hat{y}_0 - E(\hat{y}_0)}{\sqrt{\widehat{PV}(\hat{y}_0)}}$$

follows a t-distribution with (n - 2) degrees of freedom.

The $100(1-\alpha)\%$ prediction interval for $\hat{y}_0$ in this case is obtained as

$$P\left[-t_{\frac{\alpha}{2},n-2} \leq \frac{\hat{y}_0 - E(\hat{y}_0)}{\sqrt{\widehat{PV}(\hat{y}_0)}} \leq t_{\frac{\alpha}{2},n-2}\right] = 1-\alpha$$

which gives the prediction interval

$$\left[\hat{y}_0 - t_{\frac{\alpha}{2},n-2}\sqrt{MSE\left(1+\frac{1}{n}+\frac{(x_0-\overline{x})^2}{S_{xx}}\right)},\ \hat{y}_0 + t_{\frac{\alpha}{2},n-2}\sqrt{MSE\left(1+\frac{1}{n}+\frac{(x_0-\overline{x})^2}{S_{xx}}\right)}\right].$$

The prediction interval is of minimum width at $x_0 = \overline{x}$ and widens as $\left|x_0 - \overline{x}\right|$ increases.

The prediction interval for $\hat{y}_0$ is wider than the prediction interval for $\hat{\mu}_{y|x_0}$ because the prediction interval for $\hat{y}_0$ depends on both the error from the fitted model as well as the error associated with the future observations.

Reverse Regression Method

The reverse (or inverse) regression approach minimizes the sum of squares of horizontal distances between the observed data points and the line in the following scatter diagram to obtain the estimates of regression parameters.

The reverse regression has been advocated in the analysis of sex (or race) discrimination in salaries. For example, if y denotes salary and x denotes qualifications and we are interested in determining if there is a sex discrimination in salaries, we can ask:

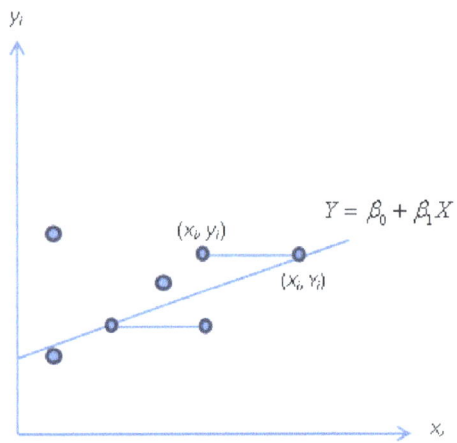

Reverse regression method

"Whether men and women with the same qualifications (value of x) are getting the same salaries (value of y). This question is answered by the direct regression."

Alternatively, we can ask:

"Whether men and women with the same salaries (value of y) have the same qualifications (value of x). This question is answered by the reverse regression, i.e., regression of x on y."

The regression equation in case of reverse regression can be written as

$$x_i = \beta_0^* + \beta_1^* y_i + \delta_i \quad (i = 1, 2, ..., n)$$

where δ_i's are the associated random error components and satisfy the assumptions as in the case of usual simple linear regression model.

The reverse regression estimates $\widehat{\beta}_{OR}$ of β_0^* and $\widehat{\beta}_{1R}$ of β_1^* for the model are obtained by interchanging the x and y in the direct regression estimators of β_0 and β_1. The estimates are obtained as

$$\widehat{\beta}_{OR} = \bar{x} - \widehat{\beta}_{1R}\,\bar{y}$$

And

$$\widehat{\beta}_{1R} = \frac{S_{xy}}{S_{yy}}$$

for β_0^* and β_1^* respectively.

The residual sum of squares in this case is

$$SS_{res}^* = s_{xx} - \frac{s_{xy}^2}{s_{yy}}.$$

Note that

$$\widehat{\beta}_{1R} b_1 = \frac{s_{xy}^2}{s_{xx} s_{yy}} = r_{xy}^2$$

where b_1 is the direct regression estimator of slope parameter and r_{xy} is the correlation coefficient between x and y. Hence if r_{xy}^2 is close to 1, the two regression lines will be close to each other.

An important application of reverse regression method is in solving the calibration problem.

Orthogonal Regression Method

The direct and reverse regression methods of estimation assume that the errors in the observations are either in x-direction or y-direction. In other words, the errors can be either in dependent variable or independent variable. There can be situations when uncertainties are involved in dependent and independent variables both. In such situations, the orthogonal regression is more appropriate. In order to take care of errors in both the directions, the least squares principle in orthogonal regression minimizes the squared perpendicular distance between the observed data points and the line in the following scatter diagram to obtain the estimates of regression coefficients. This is also known as major axis regression method. The estimates obtained are called as orthogonal regression estimates or major axis regression estimates of regression coefficients.

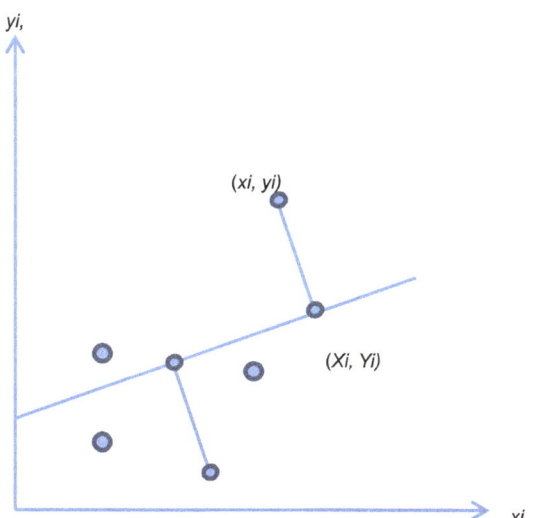

Orthogonal or major axis regression method

If we assume that the regression line to be fitted is $Y_i = \beta_0 + \beta_1 X_i$, then it is expected that all the observations (x_i, y_i), $i = 1, 2, ..., n$ lie on this line. But these points deviate from the line and in such a case, the squared perpendicular distance of observed data $(x_i, y_i)(i = 1, 2, ..., n)$ from the line is given by $d_i^2 (X_i - x_i)^2 + (Y_i - y_i)^2$ where (X_i, Y_i) denotes the ith pair of observation without any error which lie on the line.

The objective is to minimize the sum of squared perpendicular distances given by $\sum_{i=1}^{n} d_i^2$ to obtain the estimates of β_0 and β_1.

The observations $(x_i, y_i)(i = 1, 2, ..., n)$ are expected to lie on the line

$$Y_i = \beta_0 + \beta_1 X_i$$

so let

$$E_i = Y_i - \beta_0 - \beta_1 X_i = 0.$$

The regression coefficients are obtained by minimizing $\sum_{i=1}^{n} d_i^2$ under the constraints E_i's using the Lagrangian's multiplier method. The Lagrangian function is

$$L_0 = \sum_{i=1}^{n} d_i^2 - 2 \sum_{i=1}^{n} \lambda_i E_i$$

where $\lambda_i, ..., \lambda_n$ are the Lagrangian multipliers.

The set of equations are obtained by setting

$$\frac{\partial L_0}{\partial X_i} = 0, \frac{\partial L_0}{\partial Y_i} = 0, \frac{\partial L_0}{\partial \beta_0} = 0 \text{ and } \frac{\partial L_0}{\partial \beta_1} = 0 (i = 1, 2, ..., n).$$

Thus we find

$$\frac{\partial L_0}{\partial X_i} = (X_i - x_i) + \lambda_i \beta_1 = 0$$

$$\frac{\partial L_0}{\partial Y_i} = (Y_i - y_i) - \lambda_i = 0$$

$$\frac{\partial L_0}{\partial \beta_0} = \sum_{i=1}^{n} \lambda_i = 0$$

$$\frac{\partial L_0}{\partial \beta_1} = \sum_{i=1}^{n} \lambda_i X_i = 0.$$

Since

$$X_i = x_i - \lambda_i \beta_1$$
$$Y_i = y_i + \lambda_i$$

so substituting these values in E_i, we obtain

$$E_i = (y_i + \lambda_i) - \beta_0 - \beta_1(x_i - \lambda_i \beta_1) = 0$$

$$\Rightarrow \lambda_i = \frac{\beta_0 + \beta_1 x_i - y_i}{1 + \beta_1^2}.$$

Also using this λ_i in the equation $\sum_{i=1}^{n} \lambda_i = 0$, we get

$$\frac{\sum_{i=1}^{n} (\beta_0 + \beta_1 x_i - y_i)}{1 + \beta_1^2} = 0$$

and using $(X_i, x_i) + \lambda_i \beta_1 = 0$ and $\sum_{i=1}^{n} \lambda_i X_i = 0$, we get

$$\sum_{i=1}^{n} \lambda_i (x_i - \lambda_i \beta_1) = 0.$$

Substituting λ_i in this equation, we get

$$\frac{\sum_{i=1}^{n} (\beta_0 x_i + \beta_1 x_i^2 - y_i x_i)}{(1 + \beta_1^2)} - \frac{\beta_1 \sum_{i=1}^{n} (\beta_0 + \beta_1 x_i - y_i)^2}{(1 + \beta_1^2)^2} = 0. \tag{3}$$

Using λ_i in the equation and using the equation $\sum_{i=1}^{n} \lambda_i = 0$, we solve

$$\frac{\sum_{i=1}^{n} (\beta_0 + \beta_1 x_i - y_i)}{1 + \beta_1^2} = 0.$$

The solution provides an orthogonal regression estimate of β_0 as

$$\hat{\beta}_{0OR} = \bar{y} - \hat{\beta}_{1OR} \bar{x}$$

where $\hat{\beta}_{1OR}$ is an orthogonal regression estimate of β_1.

Now, substituting β_{0OR} in equation (3), we get

$$\sum_{i=1}^{n} (1 + \beta_1^2) \left[\bar{y} x_i - \beta_1 \bar{x} x_i + \beta_1 x_i^2 - x_i y_i \right] - \beta_1 \sum_{i=1}^{n} (\bar{y} - \beta_1 \bar{x} + \beta_1 x_i - y_i)^2 = 0$$

or $(1 + \beta_1^2) \sum_{i=1}^{n} x_i \left[y_i - \bar{y} - \beta_1 (x_i - \bar{x}) \right] + \beta_1 \sum_{i=1}^{n} \left[-(y_i - \bar{y}) + \beta_1 (x_i - \bar{x}) \right]^2 = 0$

or $(1 + \beta_1^2) \sum_{i=1}^{n} (u_i + \bar{x})(v_i - \beta_1 u_i) + \beta_1 \sum_{i=1}^{n} (-v_i + \beta_1 u_i)^2 = 0$

where

$$u_i = x_i - \overline{x},$$
$$v_i = y_i - \overline{y},$$

Since $\sum_{i=1}^{n} u_i = \sum_{i=1}^{n} v_i = 0$, so

$$\sum_{i=1}^{n} \left[\beta_1^2 u_i v_i + \beta_1 (u_i^2 - v_i^2) - u_i v_i \right] = 0$$

or

$$\beta_1^2 s_{xy} + \beta_1 (s_{xx} - s_{yy}) - s_{xy} = 0.$$

Solving this quadratic equation provides the orthogonal regression estimate of β_1 as

$$\hat{\beta}_{1OR} = \frac{(s_{yy} - s_{xx}) + sign(s_{xy})\sqrt{(s_{xx} - s_{yy})^2 + 4s_{xy}^2}}{2s_{xy}}$$

where sign (s_{xy}) denotes the sign of s_{xy} which can be positive or negative. So

$$sign(s_{xy}) = \begin{cases} 1 \ if \ s_{xy} > 0 \\ -1 \ if \ s_{xy} < 0. \end{cases}$$

Notice that this gives two solutions for $\hat{\beta}_{1OR}$. We choose the solution which minimizes $\sum_{i=1}^{n} d_i^2$. The other solution maximizes $\sum_{i=1}^{n} d_i^2$ and is in the direction perpendicular to the optimal solution. The optimal solution can be chosen with the sign of s_{xy}.

Reduced Major Axis Regression Method

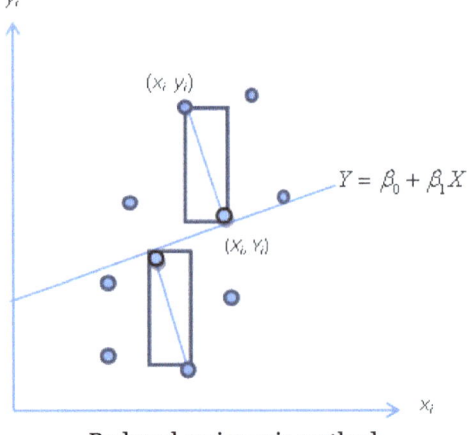

Reduced major axis method

The direct, reverse and orthogonal methods of estimation minimize the errors in a particular direction which is usually the distance between the observed data points and the line in the scatter diagram. Alternatively, one can consider the area extended by the data points in certain neighbourhood and instead of distances, the area of rectangles defined between corresponding observed data point and nearest point on the line in the following scatter diagram can also be minimized. Such an approach is more appropriate when the uncertainties are present in study as well as explanatory variables. This approach is termed as reduced major axis regression.

Suppose the regression line $Y_i = \beta_0 + \beta_1 X_i$ is on which all the observed points are expected to lie. Suppose the points $(x_i, y_i), i = 1, 2, ..., n$ are observed which lie away from the line.

The area of rectangle extended between the ith observed data point and the line is

$$A_i = (X_i \sim x_i)(Y_i \sim y_i) \quad (i = 1, 2, ..., n)$$

where $(X_i - Y_i)$ denotes the ith pair of observation without any error which lie on the line.

The total area extended by n data points is $\sum_{i=1}^{n} A_i = \sum_{i=1}^{n} (X_i \sim x_i)(Y_i \sim y_i)$.

All observed data points $(x_i, y_i), (i = 1, 2, ..., n)$ are expected to lie on the line

$$Y_i = \beta_0 + \beta_1 X_i$$

and let

$$E_i^* = Y_i - \beta_0 - \beta_1 X_i = 0.$$

So now the objective is to minimize the sum of areas under the constraints E_i^* to obtain the reduced major axis estimates of regression coefficients. Using the Lagrangian multiplies method, the Lagrangian function is

$$L_R = \sum_{i=1}^{n} A_i - \sum_{i=1}^{n} \mu_i E_i^*$$

$$= \sum_{i=1}^{n} (X_i - x_i)(Y_i - y_i) - \sum_{i=1}^{n} \mu_i E_i^*$$

where $\mu_i, ..., \mu_n$ are the Lagrangian multipliers. The set of equations are obtained by setting

$$\frac{\partial L_R}{\partial X_i} = 0, \frac{\partial L_R}{\partial Y_i} = 0, \frac{\partial L_R}{\partial \beta_0} = 0, \frac{\partial L_R}{\partial \beta_1} = 0 \quad (i = 1, 2, ..., n).$$

Thus

$$\frac{\partial L_R}{\partial X_i} = (Y_i - y_i) + \beta_1 \mu_i = 0$$

$$\frac{\partial L_R}{\partial Y_i} = (X_i - x_i) - \mu_i = 0$$

$$\frac{\partial L_R}{\partial \beta_0} = \sum_{i=1}^{n} \mu_i = 0$$

$$\frac{\partial L_R}{\partial \beta_1} = \sum_{i=1}^{n} \mu_i X_i = 0.$$

Now

$$X_i = x_i + \mu_i$$
$$Y_i = y_i - \beta_1 \mu_i$$
$$\beta_0 + \beta_1 X_i = y_i - \beta_1 \mu_i$$
$$\beta_0 + \beta_1 (x_i + \mu_i) = y_i - \beta_1 \mu_i$$
$$\Rightarrow \mu_i = \frac{y_i - \beta_0 - \beta_1 x_i}{2\beta_1}.$$

Substituting μ_i in $\sum_{i=1}^{n} \mu_i = 0$, we get the reduced major axis regression estimate of β_0 is obtained as

$$\widehat{\beta}_{0RM} = \bar{y} - \widehat{\beta}_{1RM} \bar{x}$$

where $\widehat{\beta}_{1RM}$ is the reduced major axis regression estimate of β_1. Using $X_i = x_i + \mu_i$, μ_i and $\widehat{\beta}_{0RM}$ in $\sum_{i=1}^{n} \mu_i X_i = 0$, we get

$$\sum_{i=1}^{n} \left(\frac{y_i - \bar{y} + \beta_1 \bar{x} - \beta_1 x_i}{2\beta_1} \right) \left(x_i - \frac{y_i - \bar{y} + \beta_1 \bar{x} - \beta_1 x_i}{2\beta_1} \right) = 0.$$

Let $u_i = x_i - \bar{x}$ and $v_i = y_i - \bar{x}$, then this equation can be re-expressed as $\sum_{i=1}^{n} (v_i - \beta_1 u_i)(v_i + \beta_1 u_i + 2\beta_1 \bar{x}) = 0.$

Using $\sum_{i=1}^{n} \mu_i = \sum_{i=1}^{n} v_i = 0$, we get

$$\sum_{i=1}^{n} v_i^2 = \beta_1^2 \sum_{i=1}^{n} u_i^2 = 0.$$

Solving this equation, the reduced major axis regression estimate of β_1 is obtained as

$$\widehat{\beta}_{1RM} = sign(s_{xy}) \sqrt{\frac{s_{yy}}{s_{xx}}}$$

Where

$$sign(s_{xy}) = \begin{cases} 1 \ if \ s_{xy} > 0 \\ -1 \ if \ s_{xy} < 0. \end{cases}$$

We choose the regression estimator which has same sign as that of s_{xy}.

Least Absolute Deviation Regression Method

The least squares principle advocates the minimization of sum of squared errors. The idea of squaring the errors is useful in place of simple errors because the random errors can be positive as well as negative. So consequently their sum can be close to zero indicating that there is no error in the model which can be misleading. Instead of the sum of random errors, the sum of absolute random errors can be considered which avoids the problem due to positive and negative random errors.

In the method of least squares, the estimates of the parameters β_0 and β_1 in the model $y_i = \beta_0 + \beta_1 x_i + \varepsilon_i . (i = 1, 2, ..., n)$ are chosen such that the sum of squares of deviations $\sum_{i=1}^{n} \varepsilon_i^2$ is minimum. In the method of least absolute deviation (LAD) regression, the parameters β_0 and β_1 are estimated such that the sum of absolute deviations $\sum_{i=1}^{n} |\varepsilon_i|$ is minimum. It minimizes the absolute vertical sum of errors as in the following scatter diagram:

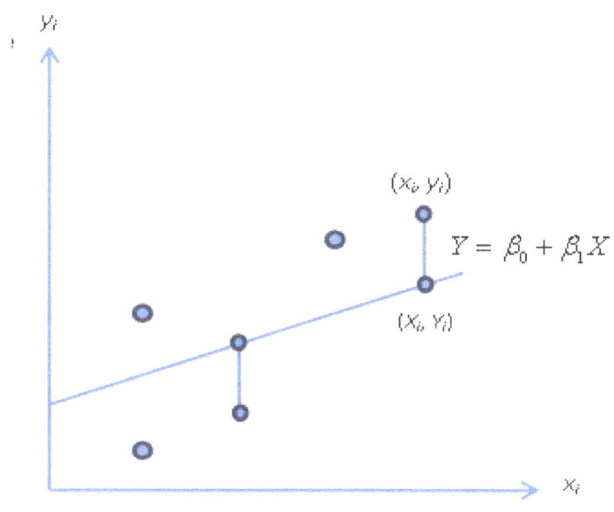

Least absolute deviation regression

The LAD estimates $\widehat{\beta}_{0L}$ and $\widehat{\beta}_{1L}$ are the values β_0 and β_1, respectively which minimize $LAD(\beta_0, \beta_1) = \sum_{i=1}^{n} |y_i - \beta_0 - \beta_1 x_i|$ for the given observations $(x_i, y_i)(i = 1, 2, ..., n)$.

Conceptually, LAD procedure is simpler than OLS procedure because |e| (absolute residuals) is a more straightforward measure of the size of the residual than e_2 (squared residuals). The LAD regression estimates of β_0 and β_1 are not available in closed form. Rather they can be obtained numerically based on algorithms. Moreover, this creates the problems of non-uniqueness and degeneracy in the estimates. The concept of non-uniqueness relates to more than one best lines passing through a data point. The degeneracy concept describes that the best line through a data point also passes through more than one other data points. The non-uniqueness and degeneracy concepts are used in algorithms to judge the quality of the estimates. The algorithm for finding the estimators generally proceeds in steps. At each step, the best line is found that passes through a given data point. The best line always passes through another data point, and this data point is used in the next step. When there is non-uniqueness, then there are more than one best lines. When there is degeneracy, then the best line passes through more than one other data point. When either of the problem is present, then there is more than one choice for the data point to be used in the next step and the algorithm may go around in circles or make a wrong choice of the LAD regression line. The exact tests of hypothesis and confidence intervals for the LAD regression estimates can not be derived analytically. Instead they are derived analogous to the tests of hypothesis and confidence intervals related to ordinary least squares estimates.

Estimation of Parameters when X is Stochastic

In a usual linear regression model, the study variable is supped to be random and explanatory variables are assumed to be fixed. In practice, there may be situations in which the explanatory variable also becomes random.

Suppose both dependent and independent variables are stochastic in the simple linear regression model

$$y = \beta_0 + \beta_1 X + \varepsilon$$

where ε is the associated random error component. The observations $(x_i, y_i), i = 1, 2, ..., n$ are assumed to be jointly distributed. Then the statistical inferences can be drawn in such cases which are conditional on X.

Assume the joint distribution of X and y to be bivariate normal $N(\mu_x, \mu_y, \sigma_x^2, \sigma_y^2, \rho)$ where μ_x and μ_y are the means of X and $y; \sigma_x^2$ and σ_y^2 are the variances of X and y, ρ and is the correlation coefficient between X and y. Then the conditional distribution of y given X = x is univariate normal conditional mean

$$E(y \mid X = x) = \mu_{y|x} = \beta_0 + \beta_1 X$$

and conditional variance of y given X = x is

$$Var(y \mid X = x) = \sigma_{y|x}^2 = \sigma_y^2(1 - \rho^2)$$

Where

$$\beta_0 = \mu_y - \mu_x \beta_1$$

and

$$\beta_1 = \frac{\sigma_y}{\sigma_x} \rho.$$

When both X and y are stochastic, then the problem of estimation of parameters can be reformulated as follows. Consider a conditional random variable y|X = x having a normal distribution with mean as conditional mean $\mu_{y|x}$ and variance as conditional variance $Var(y \mid X = x) = \sigma_{y|x}^2$. Obtain n independently distributed observation $y_i \mid x_i, i = 1, 2, ..., n$ from $N(\mu_{y|x}, \sigma_{y|x}^2)$ with nonstochastic X. Now the method of maximum likelihood can be used to estimate the parameters which yields the estimates of β_0 and β_1 as earlier in the case of nonstochastic X as

$$\tilde{b} = \bar{y} - \tilde{b}_1 \bar{x}$$

and

$$\tilde{b}_i = \frac{S_{xy}}{S_{xx}}$$

respectively.

Moreover, the correlation coefficient

$$\rho = \frac{E(y - \mu_y)(X - \mu_x)}{\sigma_x \sigma_y}$$

can be estimated by the sample correlation coefficient

$$\hat{\rho} = \frac{\sum_{i=1}^{n} E(y_i - \bar{y})(x_i - \bar{x})}{\sqrt{\sum_{i=1}^{n}(x_i - \bar{x})^2} \sqrt{\sum_{i=1}^{n}(y_i - \bar{y})^2}}$$

$$= \frac{S_{xy}}{\sqrt{S_{xx}} \sqrt{S_{yy}}}$$

$$= \tilde{b}_1 \sqrt{\frac{S_{xx}}{S_{yy}}}.$$

Thus

$$\hat{\rho}^2 = \tilde{b}_1 \frac{S_{xx}}{S_{yy}}$$

$$= \tilde{b}_1 \frac{S_{xy}}{S_{yy}}$$

$$= \frac{S_{yy} - \sum_{i=1}^{n} \hat{\varepsilon}_i^2}{S_{yy}}$$

$$= R^2$$

which is same as the coefficient of determination.

Thus R^2 has the same expression as in the case when X is fixed.

Thus R^2 again measures the goodness of fitted model even when X is stochastic.

References

- Ronald A. Fisher (1954). Statistical Methods for Research Workers (Twelfth ed.). Edinburgh: Oliver and Boyd. ISBN 0-05-002170-2

- Armstrong, J. Scott (2012). "Illusions in Regression Analysis". International Journal of Forecasting (forthcoming). 28 (3): 689. doi:10.1016/j.ijforecast.2012.02.001

- Waegeman, Willem; De Baets, Bernard; Boullart, Luc (2008). "ROC analysis in ordinal regression learning". Pattern Recognition Letters. 29: 1–9. doi:10.1016/j.patrec.2007.07.019

- Yule, G. Udny (1897). "On the Theory of Correlation". Journal of the Royal Statistical Society. Blackwell Publishing. 60 (4): 812–54. JSTOR 2979746. doi:10.2307/2979746

- Galton, Francis (1989). "Kinship and Correlation (reprinted 1989)". Statistical Science. Institute of Mathematical Statistics. 4 (2): 80–86. JSTOR 2245330. doi:10.1214/ss/1177012581

- Chiang, C.L, (2003) Statistical methods of analysis, World Scientific. ISBN 981-238-310-7 - page 274 section 9.7.4 "interpolation vs extrapolation"

- Aldrich, John (2005). "Fisher and Regression". Statistical Science. 20 (4): 401–417. JSTOR 20061201. doi:10.1214/088342305000000331

- Good, P. I.; Hardin, J. W. (2009). Common Errors in Statistics (And How to Avoid Them) (3rd ed.). Hoboken, New Jersey: Wiley. p. 211. ISBN 978-0-470-45798-6

- Tofallis, C. (2009). "Least Squares Percentage Regression". Journal of Modern Applied Statistical Methods. 7: 526–534. SSRN 1406472. doi:10.2139/ssrn.1406472

- Fotheringham, AS; Wong, DWS (1 January 1991). "The modifiable areal unit problem in multivariate statistical analysis". Environment and Planning A. 23 (7): 1025–1044. doi:10.1068/a231025

- Goldberger, Arthur S. (1964). "Classical Linear Regression". Econometric Theory. New York: John Wiley & Sons. pp. 156–212 [p. 158]. ISBN 0-471-31101-4

- Anscombe, F. J. (1948). "The Validity of Comparative Experiments". Journal of the Royal Statistical Society. Series A (General). 111 (3): 181–211. JSTOR 2984159. MR 30181. doi:10.2307/2984159

- Einicke, G.A.; Falco, G.; Malos, J.T. (May 2010). "EM Algorithm State Matrix Estimation for Navigation". IEEE Signal Processing Letters. 17 (5): 437–440. doi:10.1109/LSP.2010.2043151

- Spiegel, Murray R. (1975). Schaum's outline of theory and problems of probability and statistics. New York: McGraw-Hill. ISBN 0-585-26739-1

- Wilkinson, Leland (1999). "Statistical Methods in Psychology Journals; Guidelines and Explanations". American Psychologist. 5 (8): 594–604. CiteSeerX 10.1.1.120.4818. doi:10.1037/0003-066X.54.8.594

- Pfanzagl, Johann, with the assistance of R. Hamböker (1994). Parametric statistical theory. Walter de Gruyter, Berlin, DE. pp. 207–208. ISBN 3-11-013863-8

- Gelman, Andrew (2005). "Analysis of variance? Why it is more important than ever". The Annals of Statistics. 33: 1–53. doi:10.1214/009053604000001048

- Stigler, Stephen M. (1986). The history of statistics : the measurement of uncertainty before 1900. Cambridge, Mass: Belknap Press of Harvard University Press. ISBN 0-674-40340-1

3

An Integrated Study of Multiple Linear Regression Analysis

Multiple linear regression analysis studies the relationship between two or more variables and a dependent variable by including a linear equation so that every independent variable is associated with the value of every dependent variable. The chapter strategically encompasses and incorporates the major components and key concepts of multiple linear regression analysis, providing a complete understanding.

Multiple Linear Regression Analysis

We consider now the problem of regression when study variable depends on more than one explanatory or independent variables, called as multiple linear regression model. This model generalizes the simple linear regression in two ways. It allows the mean function E(y) to depend on more than one explanatory variables and to have shapes other than straight lines, although it does not allow for arbitrary shapes.

The Multiple Linear Regression Model

Let y denotes the dependent (or study) variable that is linearly related to k independent (or explanatory) variables $X_1, X_2, ..., X_k$ through the parameters $\beta_1, \beta_2, ..., \beta_k$ and we write

$$y = X_1\beta_1 + X_2\beta_2 + ... + X_k\beta_k + \varepsilon.$$

This is called as the multiple linear regression model. The parameters $\beta_1, \beta_2, ..., \beta_k$ are the regression coefficients associated with $X_1, X_2, ..., X_k$ respectively and ε is the random error component reflecting the difference between the observed and fitted linear relationship. There can be various reasons for such difference, e.g., joint effect of those variables not included in the model, random factors which can not be accounted in the model etc.

The jth regression coefficient β_j represents the expected change in y per unit change in jth independent variable X_j.

Assuming $E(\varepsilon) = 0$,

$$\beta^j = \frac{\partial E(y)}{\partial X_j}.$$

Linear Model

In statistics, the term linear model is used in different ways according to the context. The most common occurrence is in connection with regression models and the term is often taken as synonymous with linear regression model. However, the term is also used in time series analysis with a different meaning. In each case, the designation "linear" is used to identify a subclass of models for which substantial reduction in the complexity of the related statistical theory is possible.

Linear Regression Models

For the regression case, the statistical model is as follows. Given a (random) sample $(Y_i, X_{i1}, \ldots, X_{ip}), i = 1, \ldots, n$ the relation between the observations Y_i and the independent variables X_{ij} is formulated as

$$Y_i = \beta_0 + \beta_1\phi_1(X_{i1}) + \cdots + \beta_p\phi_p(X_{ip}) + \varepsilon_i \qquad i = 1, \ldots, n$$

where $\phi_1, \ldots, \phi_p$ may be nonlinear functions. In the above, the quantities ε_i are random variables representing errors in the relationship. The "linear" part of the designation relates to the appearance of the regression coefficients, β_j in a linear way in the above relationship. Alternatively, one may say that the predicted values corresponding to the above model, namely

$$\hat{Y}_i = \beta_0 + \beta_1\phi_1(X_{i1}) + \cdots + \beta_p\phi_p(X_{ip}) \qquad (i = 1, \ldots, n),$$

are linear functions of the β_j.

Given that estimation is undertaken on the basis of a least squares analysis, estimates of the unknown parameters β_j are determined by minimising a sum of squares function

$$S = \sum_{i=1}^{n} \left(Y_i - \beta_0 - \beta_1\phi_1(X_{i1}) - \cdots - \beta_p\phi_p(X_{ip}) \right)^2.$$

From this, it can readily be seen that the "linear" aspect of the model means the following:

- the function to be minimised is a quadratic function of the β_j for which minimisation is a relatively simple problem;

- the derivatives of the function are linear functions of the β_j making it easy to find the minimising values;

- the minimising values β_j are linear functions of the observations Y_i;

- the minimising values β_j are linear functions of the random errors ε_i which makes it relatively easy to determine the statistical properties of the estimated values of β_j.

Time Series Models

An example of a linear time series model is an autoregressive moving average model. Here the model for values $\{X_t\}$ in a time series can be written in the form

$$X_t = c + \varepsilon_t + \sum_{i=1}^{p} \phi_i X_{t-i} + \sum_{i=1}^{q} \theta_i \varepsilon_{t-i}.$$

where again the quantities ε_t are random variables representing innovations which are new random effects that appear at a certain time but also affect values of X at later times. In this instance the use of the term "linear model" refers to the structure of the above relationship in representing X_t as a linear function of past values of the same time series and of current and past values of the innovations. This particular aspect of the structure means that it is relatively simple to derive relations for the mean and covariance properties of the time series. Note that here the "linear" part of the term "linear model" is not referring to the coefficients φ_i and θ_i, as it would be in the case of a regression model, which looks structurally similar.

Other Uses in Statistics

There are some other instances where "nonlinear model" is used to contrast with a linearly structured model, although the term "linear model" is not usually applied. One example of this is nonlinear dimensionality reduction.

A model is said to be linear when it is linear in parameters. In such a case $\dfrac{\partial y}{\partial \beta_j}$ (or equivalently $\dfrac{\partial E(y)}{\partial X_j}$) should not depend on any $\beta's$. For example

i) $y = \beta_1 + \beta_2 X$ is a linear model as it is linear is parameter.

ii) $y = \beta_1 X^{\beta_2}$ can be written as

$$\log = \log \beta_1 + \beta_2 \log X$$
$$y^* = \beta_1^* + \beta_2 x^*$$

which is linear is parameter β_1^* and β_2, but nonlinear is variables $y^* = \log y, x^* = \log x$. So it is a linear model.

iii) $y = \beta_1 + \beta_2 X + \beta_3 X^2$

is linear in parameters β_1, β_2 and β_3 but it is nonlinear is variables X. So it is a linear model.

iv) $y = \beta_1 + \dfrac{\beta_2}{X - \beta_3}$

is nonlinear in parameters and variables both. So it is a nonlinear model.

v) $y = \beta_1 + \beta_2 X^{\beta_3}$

is nonlinear in parameters and variables both. So it is a nonlinear model.

vi) $y = \beta_1 + \beta_2 X + \beta_3 X^2 + \beta_4 X^3$

is a cubic polynomial model which can be written as $y = \beta_1 + \beta_2 X_2 + \beta_3 X_3 + \beta_4 X_4$ which is linear in parameters $\beta_1, \beta_2, \beta_3, \beta_4$ and linear in variables $X_2 = X, X_3 = X^2, X_4 = X^3$.. So it is a linear model.

The income and education of a person are related. It is expected that, on an average, higher level of education provides higher income. So a simple linear regression model can be expressed as

$$income = \beta_1 + \beta_2 \, education + \varepsilon.$$

Not that β_2 reflects the change in income with respect to per unit change in education and β_1 reflects the income when education is zero as it is expected that even an illiterate person can also have some income.

Further this model neglects that most people have higher income when they are older than when they are young, regardless of education. So β_2 will over-state the marginal impact of education. If age and education are positively correlated, then the regression model will associate all the observed increase in income with an increase in education. So better model is

$$income = \beta_1 + \beta_2 \, education + \beta_3 \, age + \varepsilon.$$

Usually it is observed that the income tends to rise less rapidly in the later earning years than is early years. To accommodate such possibility, we might extend the model to

$$income = \beta_1 + \beta_2 \, education + \beta_3 \, age + \beta_4 \, age^2 + \varepsilon.$$

This is how we proceed for regression modeling in real life situation. One needs to consider the experimental condition and the phenomenon before taking the decision on how many, why and how to choose the dependent and independent variables.

Model Set Up

Let an experiment be conducted n times and the data is obtained as follows:

Observation Number	Response y	Explanatory variables $X_1 \quad X_2 \, ... \, X_k$		
1	y_1	X_{11}	$X_{12} \, ...$	X_{1k}
2	y_2	X_{21}	$X_{22} \, ...$	X_{2k}
$\vdots$	$\vdots$	$\vdots$	$\vdots \quad \ddots$	$\vdots$
n	y_n	X_{n1}	$X_{n2} \, ...$	X_{nk}

Assuming that the model is

$$y = \beta_1 X_1 + \beta_2 X_2 + ... + \beta_k X_k + \varepsilon,$$

the n-tuples of observations are also assumed to follow the same model. Thus they satisfy

$$y_1 = \beta_1 x_{11} + \beta_2 x_{12} + ... + \beta_k x_{1k} + \varepsilon_1$$
$$y_2 = \beta_1 x_{21} + \beta_2 x_{22} + ... + \beta_k x_{2k} + \varepsilon_2$$
$$\vdots$$
$$y_n = \beta_1 x_{n1} + \beta_2 x_{n2} + ... + \beta_k x_{nk} + \varepsilon_n.$$

These n equations can be written as

$$
\begin{pmatrix} y_1 \\ y_1 \\ \vdots \\ y_n \end{pmatrix} = \begin{pmatrix} x_{11} & x_{12} & \cdots & x_{1k} \\ x_{21} & x_{22} & \cdots & x_{2k} \\ \vdots & \vdots & & \vdots \\ x_{n1} & x_{n2} & \cdots & x_{nk} \end{pmatrix} \begin{pmatrix} \beta_0 \\ \beta_1 \\ \vdots \\ \beta_k \end{pmatrix} + \begin{pmatrix} \varepsilon_1 \\ \varepsilon_1 \\ \vdots \\ \varepsilon_n \end{pmatrix}
$$

or

$$ y = X\beta + \varepsilon $$

where $y = (y_1, y_2, \cdots, y_n)'$ is a $n \times 1$ vector of n observation on study variable,

$$
X = \begin{pmatrix} x_{11} & x_{12} & \cdots & x_{1k} \\ x_{21} & x_{22} & \cdots & x_{2k} \\ \vdots & \vdots & & \vdots \\ x_{n1} & x_{n2} & \cdots & x_{nk} \end{pmatrix}.
$$

is a $n \times k$ matrix of n observations on each of the k explanatory variables, $\beta = (\beta_1, \beta_2, \cdots, \beta_k)'$ is a $k \times 1$ vector of regression coefficients $\varepsilon = (\varepsilon_1, \varepsilon_2, \cdots, \varepsilon_n)'$ and is a $n \times 1$ vector of random error components or disturbance term.

If intercept term is present, take first column of X to be (1,1,...,1)'. So that

$$
X = \begin{pmatrix} 1 & x_{11} & x_{12} & \cdots & x_{1k} \\ 1 & x_{21} & x_{22} & \cdots & x_{2k} \\ & \vdots & \vdots & & \vdots \\ 1 & x_{n1} & x_{n2} & \cdots & x_{nk} \end{pmatrix}.
$$

In this case, there are (k − 1) explanatory variables and one intercept term.

Assumptions in Multiple Linear Regression Model

Some assumptions are needed in the model $y = X\beta + \varepsilon$ for drawing the statistical inferences. The following assumptions are made:

(i) $E(\varepsilon) = 0$

(ii) $E(\varepsilon\varepsilon') = \sigma^2 I_n$

(iii) $Rank(X) = k$

(iv) X is a non-stochastic matrix.

(v) $\varepsilon \sim N(0, \sigma^2 I_n)$

These assumptions are used to study the statistical properties of estimator of regression coefficients. The following assumption is required to study particularly the large sample properties of the estimators

(vi) $\lim\limits_{x \to \infty}\left(\dfrac{X'X}{n}\right) = \Delta$ exists and is a non-stochastic and nonsingular matrix (with finite elements).

The explanatory variables can also be stochastic in some cases. We assume that X is non-stochastic unless stated separately.

We consider the problems of estimation and testing of hypothesis on regression coefficient vector under the stated assumption.

Estimation of Parameters

A general procedure for the estimation of regression coefficient vector is to minimize

$$\sum_{i=1}^{n} M(\varepsilon_i) = \sum_{i=1}^{n} M(y_i - x_{i1}\beta_1 - x_{i2}\beta_2 - \ldots - x_{ik}\beta_k)$$

for a suitably chosen function M.

Some examples of choice of M are

$$M(x) = |x|$$
$$M(x) = x^2$$

$$M(x) = |x|^p, \text{ in general.}$$

We consider the principle of least square which is related to $M(x) = x^2$ and method of maximum likelihood estimation for the estimation of parameters.

Principle of Ordinary Least Squares (OLS)

Let B be the set of all possible vectors β. If there is no further information, then B is k-dimensional real Euclidean space. The object is to find a vector $b' = (b_1, b_2, \ldots, b_k)$ from B that minimizes the sum of squared deviations of $\varepsilon_i 's$ i.e.,

$$S(\beta) = \sum_{i=1}^{n} \varepsilon_i^2 = \varepsilon'\varepsilon = (y = X\beta)'(y - X\beta)$$

for given y and X. A minimum will always exist as $S(\beta)$ is a real valued, convex and differentiable function. Write

$$S(\beta) = y'y + \beta'X'X\beta - 2\beta'X'y.$$

Differentiate $S(\beta)$ with respect to β

$$\frac{\partial S(\beta)}{\partial \beta} = 2X'X\beta - 2X'y$$

$$\frac{\partial^2 S(\beta)}{\partial \beta \partial \beta'} = 2X'X \ (atleast \ non-negative \ definite).$$

The normal equation is

$$\frac{\partial S(\beta)}{\partial \beta} = 0$$

$$\Rightarrow X'Xb = X'y$$

where the following result is used:

Result: If $f(z) = Z'AZ$ is a quadratic form, Z is a $m \times 1$ vector and A is any $m \times m$ symmetric matrix then $\frac{\partial}{\partial z} F(z) = 2Az.$

Since it is assumed that rank $(X) = k$ (full rank), then $X'X$ is positive definite and unique solution of normal equation is

$$b = (X'X)^{-1}X'y$$

which is termed as ordinary least squares estimator (OLSE) of β.

Since $\frac{\partial^2 S(\beta)}{\partial \beta^2}$ is at least non-negative definite, so b minimizes $S(\beta)$.

In case, X is not of full rank, then

$$b = (X'X)^{-}X'y + [I - (X'X)^{-}X'X]\omega$$

where $(X'X)^{-}$ is the generalized inverse of $X'X$ and ω is an arbitrary vector. The generalized inverse $(X'X)^{-}$ of $X'X$ satisfies

$$X'X(X'X)^{-}X'X = X'X$$
$$X(X'X)^{-}X'X = X$$
$$X'X(X'X)^{-}X' = X'.$$

Gauss–Markov Theorem

In statistics, the Gauss–Markov theorem, named after Carl Friedrich Gauss and Andrey Markov, states that in a linear regression model in which the errors have expectation zero and are uncorrelated and have equal variances, the best linear unbiased estimator (BLUE) of the coefficients is given by the ordinary least squares (OLS) estimator, provided it exists. Here "best" means giving the lowest variance of the estimate, as compared to other unbiased, linear estimators. The errors do not need to be normal, nor do they need to be independent and identically distributed (only uncorrelated with mean zero and homoscedastic with finite variance). The requirement that the estimator be unbiased cannot be dropped, since biased estimators exist with lower variance.

Statement

Suppose we have in matrix notation,

$$\underline{y} = X\underline{\beta} + \underline{\varepsilon}, \quad (\underline{y}, \underline{\varepsilon} \in \mathbb{R}^n, \underline{\beta} \in \mathbb{R}^K \text{ and } X \in \mathbb{R}^{n \times K})$$

expanding to,

$$y_i = \sum_{j=1}^{K} \beta_j X_{ij} + \varepsilon_i \quad \forall i = 1, 2, \ldots, n$$

where β_j are non-random but unobservable parameters, X_{ij} are non-random and observable (called the "explanatory variables"), ε_i are random, and so y_i are random. The random variables ε_i are called the "disturbance", "noise" or simply "error". Note that to include a constant in the model above, one can choose to introduce the constant as a variable β_{k+1} with a newly introduced last column of X being unity i.e., $X_{i(K+1)} = 1$ for all i.

The Gauss–Markov assumptions concern the set of error random variables, ε_i:

*   They have mean zero: $\mathbb{E}[\varepsilon_i] = 0$.

*   They are homoscedastic, that is all have the same finite variance: $\text{Var}(\varepsilon_i) = \sigma^2 < \infty$, and

*   Distinct error terms are uncorrelated: $\text{Cov}(\varepsilon_i, \varepsilon_j) = 0, \forall i \neq j$.

A linear estimator of β_j is a linear combination

$$\hat{\beta}_j = c_{1j} y_1 + \cdots + c_{nj} y_n$$

in which the coefficients c_{ij} are not allowed to depend on the underlying coefficients β_j, since those are not observable, but are allowed to depend on the values X_{ij}, since these data are observable. (The dependence of the coefficients on each X_{ij} is typically nonlinear; the estimator is linear in each y_i and hence in each random ε which is why this is "linear" regression.) The estimator is said to be unbiased if and only if

$$\mathbb{E}\left[\hat{\beta}_j\right] = \beta_j$$

regardless of the values of X_{ij}. Now, let $\sum_{j=1}^{K} \lambda_j \beta_j$ bsome linear combination of the coefficients. Then the mean squared error of the corresponding estimation is

$$\mathbb{E}\left[\left(\sum_{j=1}^{K} \lambda_j \left(\hat{\beta}_j - \beta_j\right)\right)^2\right],$$

in other words it is the expectation of the square of the weighted sum (across parameters) of the differences between the estimators and the corresponding parameters to be estimated. (Since we are considering the case in which all the parameter estimates are unbiased, this mean squared error is the same as the variance of the linear combination.) The best linear unbiased estimator (BLUE) of the vector β of parameters β_j is one with the smallest mean squared error for every vector λ of linear combination parameters. This is equivalent to the condition that

$$\mathrm{Var}(\tilde{\beta}) - \mathrm{Var}(\hat{\beta})$$

is a positive semi-definite matrix for every other linear unbiased estimator $\tilde{\beta}$.

The ordinary least squares estimator (OLS) is the function

$$\hat{\beta} = (X'X)^{-1} X'y$$

of y and X (where $'$ denotes the transpose of X) that minimizes the sum of squares of residuals (misprediction amounts):

$$\sum_{i=1}^{n}\left(y_i - \hat{y}_i\right)^2 = \sum_{i=1}^{n}\left(y_i - \sum_{j=1}^{K} \hat{\beta}_j X_{ij}\right)^2.$$

The theorem now states that the OLS estimator is a BLUE. The main idea of the proof is that the least-squares estimator is uncorrelated with every linear unbiased estimator of zero, i.e., with every linear combination $a_1 y_1 + \cdots + a_n y_n$ whose coefficients do not depend upon the unobservable β but whose expected value is always zero.

Proof

Let $\tilde{\beta} = Cy$ be another linear estimator of β with $C = (X'X)^{-1} X' + D$ where D is a $K \times n$ non-zero matrix. As we're restricting to *unbiased* estimators, minimum mean squared error implies minimum variance. The goal is therefore to show that such an estimator has a variance no smaller than that of $\hat{\beta}$ the OLS estimator. We calculate:

$$\begin{aligned}
\mathbb{E}[\tilde{\beta}] &= \mathbb{E}[Cy] \\
&= \mathbb{E}\left[\left((X'X)^{-1} X' + D\right)(X\beta + \varepsilon)\right] \\
&= \left((X'X)^{-1} X' + D\right) X\beta + \left((X'X)^{-1} X' + D\right) \mathbb{E}[\varepsilon] \\
&= \left((X'X)^{-1} X' + D\right) X\beta &&\mathbb{E}[\varepsilon] = 0 \\
&= (X'X)^{-1} X'X\beta + DX\beta \\
&= (I_K + DX)\beta.
\end{aligned}$$

Therefore, $\hat{\beta}$ is unbiased if and only if $DX = 0$. Then:

$$
\begin{aligned}
(\tilde{\beta}) &= \text{Var}(Cy) \\
&= C \ \text{Var}(y)C' \\
&= \sigma^2 CC' = \sigma^2 \left((X'X)^{-1} X' + D \right) \left(X(X'X)^{-1} + D' \right) \\
&= \sigma^2 \left((X'X)^{-1} X'X(X'X)^{-1} + (X'X)^{-1} X'D' + DX(X'X)^{-1} + DD' \right) \\
&= \sigma^2 (X'X)^{-1} + \sigma^2 (X'X)^{-1}(DX)' \quad \sigma^2 DX(X'X)^{-1} + \sigma^2 DD' \\
&= \sigma^2 (X'X)^{-1} + \sigma^2 DD' \qquad\qquad\qquad\qquad DX = 0 \\
&= \text{Var}(\hat{\beta}) + \sigma^2 DD' \qquad\qquad\qquad\quad \sigma^2 (X'X)^{-1} = \text{Var}(\hat{\beta})
\end{aligned}
$$

Since DD' is a positive semidefinite matrix, $\text{Var}(\tilde{\beta})$ exceeds $\text{Var}(\tilde{\beta})$ by a positive semidefinite matrix.

Remarks on the Proof

As it has been stated before, the condition of $\text{Var}(\tilde{\beta}) - \text{Var}(\hat{\beta})$ is equivalent to the property that the best linear unbiased estimator of $l'\beta$ is $l'\hat{\beta}$ (best in the sense that it has minimum variance). To see this, let $l'\tilde{\beta}$ another linear unbiased estimator of $l'\beta$.

$$
\begin{aligned}
\text{Var}(l'\tilde{\beta}) &= l'\text{Var}(\tilde{\beta})l \\
&= \sigma^2 l'(X'X)^{-1} l + l'DD'l \\
&= \text{Var}(l'\hat{\beta}) + (D'l)'(D'l) \qquad\qquad \sigma^2 l'(X'X)^{-1} l = \text{Var}(l'\hat{\beta}) \\
&= \text{Var}(l'\hat{\beta}) + \| D'l \| \\
&\geqslant \text{Var}(l'\hat{\beta})
\end{aligned}
$$

Moreover equality holds if and only if $D'l = 0$. We calculate

$$
\begin{aligned}
l'\tilde{\beta} &= l'\left(((X'X)^{-1} X' + D)Y \right) \qquad\qquad \text{from above} \\
&= l'(X'X)^{-1} X'Y + l'DY \\
&= l'\hat{\beta} + (D'l)'Y \\
&= l'\hat{\beta} \qquad\qquad\qquad\qquad\qquad D'l = 0
\end{aligned}
$$

This proves that the equality holds if and only if $l'\tilde{\beta} = l'\hat{\beta}$ which gives the uniqueness of the OLS estimator as a BLUE.

Generalized Least Squares Estimator

The generalized least squares (GLS), developed by Aitken, extends the Gauss–Markov theorem to the case where the error vector has a non-scalar covariance matrix. The Aitken estimator is also a BLUE.

Gauss–Markov Theorem as Stated in Econometrics

In most treatments of OLS, the regressors in the design matrix $\mathbf{X}$ are assumed to be fixed in repeated samples. This assumption is considered inappropriate for a predominantly nonexperimental science like econometrics. Instead, the assumptions of the Gauss–Markov theorem are stated conditional on $\mathbf{X}$.

Linearity

The dependent variable is assumed to be a linear function of the variables specified in the model. The specification must be linear in its parameters. This does not mean that there must be a linear relationship between the independent and dependent variables. The independent variables can take non-linear forms as long as the parameters are linear. The equation $y = \beta_0 + \beta_1 x^2$, qualifies as linear while $y = \beta_0 + \beta_1^2 x$ can be transformed to be linear by replacing β_1^2 by another parameter, say γ. An equation with a parameter dependent on an independent variable does not qualify as linear, for example $y = \beta_0 + \beta_1(x) \cdot x$, where $\beta_1(x)$ is a function of x.

Data transformations are often used to convert an equation into a linear form. For example, the Cobb–Douglas function—often used in economics—is nonlinear:

$$Y = AL^\alpha K^{1-\alpha} e^\varepsilon$$

But it can be expressed in linear form by taking the natural logarithm of both sides:

$$\ln Y = \ln A + \alpha \ln L + (1-\alpha)\ln K + \varepsilon = \beta_0 + \beta_1 \ln L + \beta_2 \ln K + \varepsilon$$

This assumption also covers specification issues: assuming that the proper functional form has been selected and there are no omitted variables.

Strict Exogeneity

For all n observations, the expectation—conditional on the regressors—of the error term is zero:

$$\mathrm{E}[\varepsilon_i|\mathbf{X}] = \mathrm{E}[\varepsilon_i|\mathbf{x_1},\ldots,\mathbf{x_n}] = 0.$$

where $\mathbf{x}_i = \begin{bmatrix} x_{i1} & x_{i2} & \ldots & x_{ik} \end{bmatrix}^\mathsf{T}$ is the data vector of regressors for the ith observation, and consequently $\mathbf{X} = \begin{bmatrix} \mathbf{x_1^\mathsf{T}} & \mathbf{x_2^\mathsf{T}} & \ldots & \mathbf{x_n^\mathsf{T}} \end{bmatrix}^\mathsf{T}$ is the data matrix or design matrix.

Geometrically, this assumptions implies that $\mathbf{x}_i$ and ε_i are orthogonal to each other, so that their inner product (i.e., their cross moment) is zero.

$$\mathrm{E}[\mathbf{x}_j \cdot \varepsilon_i] = \begin{bmatrix} \mathrm{E}[x_{j1} \cdot \varepsilon_i] \\ \mathrm{E}[x_{j2} \cdot \varepsilon_i] \\ \vdots \\ \mathrm{E}[x_{jk} \cdot \varepsilon_i] \end{bmatrix} = \mathbf{0} \quad \text{for all } i, j \in n$$

This assumption is violated if the explanatory variables are stochastic, for instance when they are measured with error, or are endogenous. Endogeneity can be the result of simultaneity, where causality flows back and forth between both the dependent and independent variable. Instrumental variable techniques are commonly used to address this problem.

Full Rank

The sample data matrix $\mathbf{X}$ must be non-singular, i.e. it must have full rank.

$$\text{rank}(\mathbf{X}) = k$$

Otherwise $\mathbf{X}$ is not invertible and the OLS estimator cannot be computed.

A violation of this assumption is perfect multicollinearity, i.e. some explanatory variables are linearly dependent. One scenario in which this will occur is called "dummy variable trap," when a base dummy variable is not omitted resulting in perfect correlation between the dummy variables and the constant term.

Multicollinearity (as long as it is not "perfect") can be present resulting in a less efficient, but still unbiased estimate. The estimates will be less precise and highly sensitive to particular sets of data. Multicollinearity can be detected from condition number or the variance inflation factor, among other tests.

Spherical Errors

The outer product of the error vector must be spherical.

$$\mathrm{E}[\varepsilon\varepsilon^T | X] = Var[\varepsilon | X] = \begin{bmatrix} \sigma^2 & 0 & \cdots & 0 \\ 0 & \sigma^2 & \cdots & 0 \\ \vdots & \vdots & \ddots & \vdots \\ 0 & 0 & \cdots & \sigma^2 \end{bmatrix} = \sigma^2 I \quad \text{with } \sigma^2 > 0$$

This implies the error term has uniform variance (homoscedasticity) and no serial dependence. If this assumption is violated, OLS is still unbiased, but inefficient. The term "spherical errors" will describe the multivariate normal distribution: if $Var[\varepsilon | X] = \sigma^2 I$ in the multivariate normal density, then the equation f(x)=c is the formula for a "ball" centered at μ with radius σ in n-dimensional space.

Heteroskedasticity occurs when the amount of error is correlated with an independent variable. For example, in a regression on food expenditure and income, the error is correlated with income. Low income people generally spend a similar amount on food, while high income people may spend a very large amount or as little as low income people spend. Heteroskedastic can also be caused by changes in measurement practices. For example, as statistical offices improve their data, measurement error decreases, so the error term declines over time.

This assumption is violated when there is autocorrelation. Autocorrelation can be visualized on a data plot when a given observation is more likely to lie above a fitted line if adjacent observations also lie above the fitted regression line. Autocorrelation is common in time series data where a data series may experience "inertia." If a dependent variable takes a while to fully absorb a shock. Spatial autocorrelation can also occur geographic areas are likely to have similar errors. Autocorrelation may be the result of misspecification such as choosing the wrong functional form. In these cases, correcting the specification is one possible way to deal with autocorrelation.

In the presence of non-spherical errors, the generalized least squares estimator can be shown to be blue.

Theorem

i. Let $\hat{y} = Xb$ be the empirical predictor of y. Then $\hat{y}$ has the same value for all solutions b of $X'Xb = X'y$

ii. $S(\beta)$ attains the minimum for any solution of $X'Xb = X'y$.

Proof:

i. Let b be any member in $b = (X'X)^- X'y + \left[I - (X'X)^- X'X \right] \omega$

Since $X(X'X)^- X'X = X$, so then

$$Xb = X(X'X)^- X'y + X\left[I - (X'X)^- X'X \right] \omega$$
$$= X(X'X)^- X'y$$

which is independent of ω. This implies that $\hat{y}$ has same value for all solution b of $X'Xb = X'y$.

ii. Note that for any β,

$$\begin{aligned}
S(\beta) &= \left[y - Xb + X(b - \beta) \right]' \left[y - Xb + X(b - \beta) \right] \\
&= (y - Xb)'(y - Xb) + (b - \beta)'X'X(b - \beta) + 2(b - \beta)'X'(y - Xb) \\
&= (y - Xb)'(y - Xb) + (b - \beta)'X'X(b - \beta) \quad \text{(Using } X'Xb = X'y) \\
&\geq (y - Xb)'(y - Xb) = S(b) \\
&= y'y - 2y'Xb + b'X'Xb \\
&= y'y - b'X'Xb \\
&= y'y - \hat{y}'\hat{y}.
\end{aligned}$$

Fitted Values

Now onwards, we assume that X is a full column rank matrix.

If $\hat{\beta}$ is any estimator of β for the model $y = X\beta + \varepsilon$, then the fitted values are defined as $\hat{y} = X\hat{\beta}$ where $\hat{\beta}$ is any estimator of β.

In case of $\hat{\beta} = b$,

$$\hat{y} = Xb$$
$$= X(X'X)^{-1}X'y$$
$$= Hy$$

where $H = X(X'X)^{-1}X'$ is termed as Hat Matrix which is

i. symmetric

ii. idempotent (i.e., $HH = H$) and

iii. $tr\, H = trX(X'X)^{-1}X' = tr\, X'X(X'X)^{-1} = tr\, I_k = k$.

Residuals

The difference between the observed and fitted values of study variable is called as residual. It is denoted as

$$e = y \sim \hat{y}$$
$$= y - \hat{y}$$
$$= y - Xb$$
$$= y - Hy$$
$$= (I - H)y$$
$$= \overline{H}y$$

where $\overline{H} = I - H$.

Note that

(i) $\overline{H}$ is a symmetric matrix,

(ii) $\overline{H}$ is an idempotent matrix, i.e., $\overline{H}\,\overline{H} = (I - H)(I - H) = (I - H) = \overline{H}$ and

(iii) $tr\overline{H} = trI_n - trII = (n - k)$.

Properties of OLSE

(i) Estimation error

The estimation error of b is

$$b - \beta = (X'X)^{-1}X'y - \beta$$
$$= (X'X)^{-1}X'(X\beta + \varepsilon) - \beta$$
$$= (X'X)^{-1}X'\varepsilon.$$

(ii) Bias

Since X is assumed to be nonstochastic and $E(\varepsilon) = 0$

$$E(b\text{-}\beta) = (X'X)^{-1}X'(\varepsilon)$$
$$= 0.$$

Thus OLSE is an unbiased estimator of β.

(iii) Covariance matrix

The covariance matrix of b is

$$V(b) = E(b\text{-}\beta)(b\text{-}\beta)'$$
$$= E\left[(X'X)^{-1}X'E\varepsilon\varepsilon'X(X'X)^{-1}\right]$$
$$= (X'X)^{-1}X'E(\varepsilon\varepsilon')X(X'X)^{-1}$$
$$= \sigma^2(X'X)^{-1}X'IX(X'X)^{-1}$$
$$= \sigma^2(X'X)^{-1}.$$

(iv) Variance

The variance of b can be obtained as the sum of variances of all $b_1, b_2, ..., b_k$ which is the trace of covariance matrix of b. Thus

$$Var(b) = tr\left[V(b)\right]$$
$$= \sum_{i=1}^{k} E(b_i - \beta_i)^2$$
$$= \sum_{i=1}^{k} Var(b_i).$$

Estimation of σ²

The least squares criterion can not be used to estimate σ^2 because σ^2 does not appear in $S(\beta)$. Since $E(\varepsilon_i^2) = \sigma^2$, so we attempt with residuals e_i to estimate σ^2 as follows:

$$e = y - \hat{y}$$
$$= y - X(X'X)^{-1}X'y$$
$$= [I - X(X'X)^{-1}X']y$$
$$= \bar{H}y.$$

Consider the residual sum of squares

$$SS_{res} = \sum_{i=1}^{n} e_i^2$$
$$= e'e$$
$$= (y - Xb)'(y - Xb)$$
$$= y'(I - H)(I - H)y$$
$$= y'(I - H)y$$
$$= y'\overline{H}y.$$

Also

$$SS_{res} = (y - Xb)'(y - Xb)$$
$$= y'y - 2b'X'y + b'X'Xb$$
$$= y'y - b'X'y \qquad (\text{Using } X'Xb = X'y).$$
$$SS_{res} = y'\overline{H}y$$
$$= (X\beta + \varepsilon)'\overline{H}(X\beta + \varepsilon)$$
$$= \varepsilon'\overline{H}\varepsilon \quad (\text{Using } \overline{H}X = 0).$$

Since $\varepsilon \sim N(0, \sigma^2 I)$,

So $y \sim N(X\beta, \sigma^2 I)$.

Hence $y'\overline{H}y \sim \chi^2(n - k)$.

Thus $E\left[y'\overline{H}y\right] = (n - k)\sigma^2$

Or $E\left[\dfrac{y'\overline{H}y}{n - k}\right] = \sigma^2$

Or $E\left[MS_{res}\right] = \sigma^2$

where $MS_{res} = \dfrac{SS_{res}}{n - k}$ is the mean sum of squares due to residual.

Thus an unbiased estimator of σ^2 is

$$\hat{\sigma}^2 = MS_{res} = s^2 (\text{say}),$$

which is a model dependent estimator.

Covariance Matrix of $\hat{y}$

The covariance Matrix of $\hat{y}$ is

$$V(\hat{y}) = V(Xb)$$
$$= XV(b)X'$$
$$= \sigma^2 X(X'X)^{-1}X'$$
$$= \sigma^2 H.$$

The ordinary least squares estimator (OLSE) is the best linear unbiased estimator (BLUE) of β.

Proof: The OLSE of β is

$$b = (X'X)^{-1}X'y$$

which is a linear function of y. Consider the arbitrary linear estimator $b^* = a'y$ of linear parametric function $\ell'\beta$ where the elements of α are arbitrary constants.

Then for b^*,

$$E(b^*) = E(a'y) = a'X\beta$$

and so b^* is an unbiased estimator of $\ell'\beta$ when

$$E(b^*) = a'X\beta = \ell'\beta$$
$$\Rightarrow a'X = \ell'.$$

Since we wish to consider only those estimators that are linear and unbiased, so we restrict ourselves to those estimators for which $a'X = \ell'$.

Further

$$Var(a'y) = a'Var(y)a = \sigma^2 a'a$$
$$Var(\ell'b) = \ell'Var(b)\ell$$
$$= \sigma^2 a'X(X'X)^{-1}a.$$

Consider

$$Var(a'y) - Var(\ell'b) = \sigma^2\left[a'a - a'X(X'X)^{-1}X'a\right]$$
$$= \sigma^2 a'\left[I - X(X'X)^{-1}X'\right]a$$
$$= \sigma^2 a'(I - H)a.$$

Since (I - H) is a positive semi-definite matrix, so

$$\text{Var(a'y)} - Var(\ell'b) \geq 0.$$

This reveals that if b^* is any linear unbiased estimator then its variance must be no smaller than that of b.

If we consider $\ell = (0,0,...,0,1,0,...,0)$ (here 1 occurs at i[th] place), then $\ell'b = b_i$ is best linear unbiased estimator of $\ell'\beta = \beta_i$ for all i = 1,2,...,k.

Consequently b is the best linear unbiased estimator of β, where 'best' refers to the fact that b is efficient within the class of linear and unbiased estimators.

Cramér–Rao Bound

In estimation theory and statistics, the Cramér–Rao bound (CRB) or Cramér–Rao lower bound (CRLB), named in honor of Harald Cramér and Calyampudi Radhakrishna Rao who were among the first to derive it, expresses a lower bound on the variance of estimators of a deterministic (fixed, though unknown) parameter. The bound is also known as the Cramér–Rao inequality or the information inequality.

In its simplest form, the bound states that the variance of any unbiased estimator is at least as high as the inverse of the Fisher information. An unbiased estimator which achieves this lower bound is said to be (fully) efficient. Such a solution achieves the lowest possible mean squared error among all unbiased methods, and is therefore the minimum variance unbiased (MVU) estimator. However, in some cases, no unbiased technique exists which achieves the bound. This may occur even when an MVU estimator exists.

The Cramér–Rao bound can also be used to bound the variance of *biased* estimators of given bias. In some cases, a biased approach can result in both a variance and a mean squared error that are *below* the unbiased Cramér–Rao lower bound.

Statement

The Cramer–Rao bound is stated in this section for several increasingly general cases, beginning with the case in which the parameter is a scalar and its estimator is unbiased. All versions of the bound require certain regularity conditions, which hold for most well-behaved distributions.

Scalar Unbiased Case

Suppose θ is an unknown deterministic parameter which is to be estimated from measurements x, distributed according to some probability density function $f(x;\theta)$. The variance of any *unbiased* estimator $\hat{\theta}$ of θ is then bounded by the reciprocal of the Fisher information $I(\theta)$:

$$var(\hat{\theta}) \geq \frac{1}{I(\theta)}$$

where the Fisher information $I(\theta)$ is defined by

$$I(\theta) = E\left[\left(\frac{\partial \ell(x;\theta)}{\partial \theta}\right)^2\right] = -E\left[\frac{\partial^2 \ell(x;\theta)}{\partial \theta^2}\right]$$

and $\ell(x;\theta) = \log(f(x;\theta))$ is the natural logarithm of the likelihood function and E denotes the expected value (over x).

The efficiency of an unbiased estimator $\hat{\theta}$ measures how close this estimator's variance comes to this lower bound; estimator efficiency is defined as

$$e(\hat{\theta}) = \frac{I(\theta)^{-1}}{var(\hat{\theta})}$$

or the minimum possible variance for an unbiased estimator divided by its actual variance. The Cramér–Rao lower bound thus gives

$$e(\hat{\theta}) \leq 1.$$

General Scalar Case

A more general form of the bound can be obtained by considering an unbiased estimator $T(X)$ of the parameter θ. Here, unbiasedness is understood as stating that $E\{T(X)\} = \psi(\theta)$. In this case, the bound is given by

$$var(T) \geq \frac{[\psi'(\theta)]^2}{I(\theta)}$$

where $\psi'(\theta)$ is the derivative of $\psi(\theta)$ (by θ), and $I(\theta)$ is the Fisher information defined above.

Bound on the Variance of Biased Estimators

Apart from being a bound on estimators of functions of the parameter, this approach can be used to derive a bound on the variance of biased estimators with a given bias, as follows. Consider an estimator $\hat{\theta}$ with bias $b(\theta) = E\{\hat{\theta}\} - \theta$, and let $\psi(\theta) = b(\theta) + \theta$. By the result above, any unbiased estimator whose expectation is $\psi(\theta)$ has variance greater than or equal to $(\psi'(\theta))^2 / I(\theta)$. Thus, any estimator $\hat{\theta}$ whose bias is given by a function $b(\theta)$ satisfies

$$Var\left(\hat{\theta}\right) \geq \frac{[1+b'(\theta)]^2}{I(\theta)}.$$

The unbiased version of the bound is a special case of this result, with $b(\theta) = 0$.

It's trivial to have a small variance – an "estimator" that is constant has a variance of zero. But from the above equation we find that the mean squared error of a biased estimator is bounded by

$$E\left((\hat{\theta} - \theta)^2\right) \geq \frac{[1+b'(\theta)]^2}{I(\theta)} + b(\theta)^2,$$

using the standard decomposition of the MSE. Note, however, that if $1 + b'(\theta) < 1$ this bound might be less than the unbiased Cramér–Rao bound $1 / I(\theta)$. For instance, in the example of estimating variance below, $1 + b'(\theta) = \dfrac{n}{n+2} < 1$.

Multivariate Case

Extending the Cramér–Rao bound to multiple parameters, define a parameter column vector

$$\theta = \left[\theta_1, \theta_2, \ldots, \theta_d\right]^T \in \mathbb{R}^d$$

with probability density function $f(x;\theta)$ which satisfies the two regularity conditions below.

The Fisher information matrix is a $d \times d$ matrix with element $I_{m,k}$ defined as

$$_{1,k} = E\left[\frac{\partial}{\partial \theta_m} \log f\left(x;\theta\right) \frac{\partial}{\partial \theta_k} \log f\left(x;\theta\right)\right] = -E\left[\frac{\partial^2}{\partial \theta_m \partial \theta_k} \log f\left(x;\theta\right)\right].$$

Let $\mathbf{T}(X)$ be an estimator of any vector function of parameters, $\mathbf{T}(X) = (T_1(X), \ldots, T_d(X))^T$, and denote its expectation vector $E[\mathbf{T}(X)]$ by $\psi(\theta)$. The Cramér–Rao bound then states that the co-variance matrix of $\mathbf{T}(X)$ satisfies

$$cov_\theta\left(T(X)\right) \geq \frac{\partial \psi\left(\theta\right)}{\partial \theta} [I\left(\theta\right)]^{-1} \left(\frac{\partial \psi\left(\theta\right)}{\partial \theta}\right)^T$$

where

- The matrix inequality $A \geq B$ is understood to mean that the matrix $A - B$ is positive semidefinite, and

- $\partial \psi(\theta) / \partial \theta$ is the Jacobian matrix whose ij element is given by $\partial \psi_i(\theta) / \partial \theta_j$.

If $\mathbf{T}(X)$ is an unbiased estimator of θ (i.e., $\psi(\theta) = \theta$), then the Cramér–Rao bound reduces to

$$cov_\theta\left(T(X)\right) \geq I\left(\theta\right)^{-1}.$$

If it is inconvenient to compute the inverse of the Fisher information matrix, then one can simply take the reciprocal of the corresponding diagonal element to find a (possibly loose) lower bound.

$$var_\theta\left(T_m(X)\right) = \left[cov_\theta\left(T(X)\right)\right]_{mm} \geq \left[I\left(\theta\right)^{-1}\right]_{mm} \geq \left(\left[I\left(\theta\right)\right]_{mm}\right)^{-1}.$$

Regularity Conditions

The bound relies on two weak regularity conditions on the probability density function, $f(x;\theta)$, and the estimator $T(X)$:

- The Fisher information is always defined; equivalently, for all x such that $f(x;\theta) > 0$,

$$\frac{\partial}{\partial \theta} \log f(x; \theta)$$

exists, and is finite.

- The operations of integration with respect to x and differentiation with respect to θ can be interchanged in the expectation of T; that is,

$$\frac{\partial}{\partial \theta}\left[\int T(x)f(x;\theta)dx\right] = \int T(x)\left[\frac{\partial}{\partial \theta}f(x;\theta)\right]dx$$

whenever the right-hand side is finite.

This condition can often be confirmed by using the fact that integration and differentiation can be swapped when either of the following cases hold:

1. The function $f(x;\theta)$ has bounded support in x, and the bounds do not depend on θ;

2. The function $f(x;\theta)$ has infinite support, is continuously differentiable, and the integral converges uniformly for all θ.

Simplified form of the Fisher Information

Suppose, in addition, that the operations of integration and differentiation can be swapped for the second derivative of $f(x;\theta)$ as well, i.e.,

$$\frac{\partial^2}{\partial \theta^2}\left[\int T(x)f(x;\theta)dx\right] = \int T(x)\left[\frac{\partial^2}{\partial \theta^2}f(x;\theta)\right]dx.$$

In this case, it can be shown that the Fisher information equals

$$I(\theta) = -E\left[\frac{\partial^2}{\partial \theta^2}\log f(X;\theta)\right].$$

The Cramèr–Rao bound can then be written as

$$var\left(\hat{\theta}\right) \geq \frac{1}{I(\theta)} = \frac{1}{-E\left[\frac{\partial^2}{\partial \theta^2}\log f(X;\theta)\right]}.$$

In some cases, this formula gives a more convenient technique for evaluating the bound.

Single-parameter Proof

The following is a proof of the general scalar case of the Cramér–Rao bound described above. Assume that $T = t(X)$ is an unbiased estimator for the value $\psi(\theta)$ (based on the observations X), and so $E(T) = \psi(\theta)$. The goal is to prove that, for all θ,

$$var(t(X)) \geq \frac{[\psi'(\theta)]^2}{I(\theta)}.$$

Let X be a random variable with probability density function $f(x;\theta)$. Here $T = t(X)$ is a statistic, which is used as an estimator for $\psi(\theta)$. Define V as the score:

$$V = \frac{\partial}{\partial\theta}\ln f(X;\theta) = \frac{1}{f(X;\theta)}\frac{\partial}{\partial\theta}f(X;\theta)$$

where the chain rule is used in the final equality above. Then the expectation of V, written $E(V)$, is zero. This is because:

$$E(V) = \int f(x;\theta)\left[\frac{1}{f(x;\theta)}\frac{\partial}{\partial\theta}f(x;\theta)\right]dx = \frac{\partial}{\partial\theta}\int f(x;\theta)dx = 0$$

where the integral and partial derivative have been interchanged (justified by the second regularity condition).

If we consider the covariance $cov(V,T)$ of V and T, we have $cov(V,T) = E(VT)$, because $E(V) = 0$. Expanding this expression we have

$$cov(V,T) = E\left(T\cdot\left[\frac{1}{f(X;\theta)}\frac{\partial}{\partial\theta}f(X;\theta)\right]\right) = \int t(x)\left[\frac{1}{f(x;\theta)}\frac{\partial}{\partial\theta}f(x;\theta)\right]f(x;\theta)dx = \frac{\partial}{\partial\theta}\left[\int t(x)f(x;\theta)dx\right] = \psi'(\theta)$$

again because the integration and differentiation operations commute (second condition).

The Cauchy–Schwarz inequality shows that

$$\sqrt{var(T)var(V)} \geq |cov(V,T)| = |\psi'(\theta)|$$

therefore

$$var(T) \geq \frac{[\psi'(\theta)]^2}{var(V)} = \frac{[\psi'(\theta)]^2}{I(\theta)}$$

which proves the proposition.

Examples

Multivariate Normal Distribution

For the case of a d-variate normal distribution

$$x \sim N_d(\mu(\theta), C(\theta))$$

the Fisher information matrix has elements

$$I_{m,k} = \frac{\partial \mathbf{\hat{i}}}{\partial \theta_m}^T \mathbf{C}^{-1} \frac{\partial \mathbf{\hat{i}}}{\partial \theta_k} + \frac{1}{2} \mathrm{tr}\left(\mathbf{C}^{-1} \frac{\partial \mathbf{C}}{\partial \theta_m} \mathbf{C}^{-1} \frac{\partial \mathbf{C}}{\partial \theta_k} \right)$$

where "tr" is the trace.

For example, let $w[n]$ be a sample of N independent observations with unknown mean θ and known variance σ^2.

$$w[n] \sim \mathbb{N}_N \left(\theta \mathbf{1}, \sigma^2 \mathbf{I} \right).$$

Then the Fisher information is a scalar given by

$$I(\theta) = \left(\frac{\partial \mathbf{\hat{i}}(\theta)}{\partial \theta} \right)^T \mathbf{C}^{-1} \left(\frac{\partial \mathbf{\hat{i}}(\theta)}{\partial \theta} \right) = \sum_{i=1}^{N} \frac{1}{\sigma^2} = \frac{N}{\sigma^2},$$

and so the Cramér–Rao bound is

$$var\left(\hat{\theta} \right) \geq \frac{\sigma^2}{N}.$$

Normal Variance with Known Mean

Suppose X is a normally distributed random variable with known mean μ and unknown variance σ^2. Consider the following statistic:

$$T = \frac{\sum_{i=1}^{n} \left(X_i - \mu \right)^2}{n}.$$

Then T is unbiased for σ^2, as $E(T) = \sigma^2$. What is the variance of T?

$$var(T) = \frac{var(X - \mu)^2}{n} = \frac{1}{n} \left[E\left\{ (X - \mu)^4 \right\} - \left(E\left\{ (X - \mu)^2 \right\} \right)^2 \right]$$

(the second equality follows directly from the definition of variance). The first term is the fourth moment about the mean and has value $3(\sigma^2)^2$; the second is the square of the variance, or $(\sigma^2)^2$. Thus

$$var(T) = \frac{2(\sigma^2)^2}{n}.$$

Now, what is the Fisher information in the sample? Recall that the score V is defined as

$$V = \frac{\partial}{\partial \sigma^2} \log L(\sigma^2, X)$$

where L is the likelihood function. Thus in this case,

$$V = \frac{\partial}{\partial \sigma^2} \log \left[\frac{1}{\sqrt{2\pi\sigma^2}} e^{-(X-\mu)^2/2\sigma^2} \right] = \frac{(X-\mu)^2}{2(\sigma^2)^2} - \frac{1}{2\sigma^2}$$

where the second equality is from elementary calculus. Thus, the information in a single observation is just minus the expectation of the derivative of V, or

$$I = -E\left(\frac{\partial V}{\partial \sigma^2}\right) = -E\left(-\frac{(X-\mu)^2}{(\sigma^2)^3} + \frac{1}{2(\sigma^2)^2}\right) = \frac{\sigma^2}{(\sigma^2)^3} - \frac{1}{2(\sigma^2)^2} = \frac{1}{2(\sigma^2)^2}.$$

Thus the information in a sample of n independent observations is just n times this, or $\dfrac{n}{2(\sigma^2)^2}$.

The Cramer Rao bound states that

$$var(T) \geq \frac{1}{I}.$$

In this case, the inequality is saturated (equality is achieved), showing that the estimator is efficient.

However, we can achieve a lower mean squared error using a biased estimator. The estimator

$$T = \frac{\sum_{i=1}^{n}(X_i - \mu)^2}{n+2}.$$

obviously has a smaller variance, which is in fact

$$var(T) = \frac{2n(\sigma^2)^2}{(n+2)^2}.$$

Its bias is

$$\left(1 - \frac{n}{n+2}\right)\sigma^2 = \frac{2\sigma^2}{n+2}$$

so its mean squared error is

$$MSE(T) = \left(\frac{2n}{(n+2)^2} + \frac{4}{(n+2)^2}\right)(\sigma^2)^2 = \frac{2(\sigma^2)^2}{n+2}$$

which is clearly less than the Cramér–Rao bound found above.

When the mean is not known, the minimum mean squared error estimate of the variance of a sample from Gaussian distribution is achieved by dividing by $n + 1$, rather than $n - 1$ or $n + 2$.

Let $\theta = (\beta, \sigma^2)'$. Assume that both β and σ^2 are unknown. If $E(\hat{\theta}) = \theta$, then the Cramer-Rao lower bound for $\hat{\theta}$ is greater than or equal to the matrix inverse of

$$I(\theta) = -E\left[\frac{\partial^2 \ln L(\theta)}{\partial \theta \partial \theta'}\right]$$

$$= \begin{bmatrix} -E\left[\dfrac{\partial \ln L(\beta, \sigma^2)}{\partial \beta^2}\right] & -E\left[\dfrac{\partial \ln L(\beta, \sigma^2)}{\partial \beta \partial \sigma^2}\right] \\[2ex] -E\left[\dfrac{\partial \ln L(\beta, \sigma^2)}{\partial \sigma^2 \partial \beta}\right] & -E\left[\dfrac{\partial \ln L(\beta, \sigma^2)}{\partial^2 (\sigma^2)^2}\right] \end{bmatrix}$$

$$= \begin{bmatrix} -E\left[-\dfrac{X'X}{\sigma^2}\right] & -E\left[\dfrac{X'(y-X\beta)}{\sigma^4}\right] \\[2ex] -E\left[\dfrac{(y-X\beta)'X}{\sigma^4}\right] & -E\left[\dfrac{n}{2\sigma^4} - \dfrac{(y-X\beta)'(y-X\beta)}{\sigma^6}\right] \end{bmatrix}$$

$$= \begin{bmatrix} \dfrac{X'X}{\sigma^2} & 0 \\[2ex] 0 & \dfrac{n}{2\sigma^4} \end{bmatrix}.$$

Then

$$[I(\theta)]^{-1} = \begin{bmatrix} \sigma^2 (X'X)^{-1} & 0 \\[2ex] 0 & \dfrac{2\sigma^4}{n} \end{bmatrix}$$

is the Cramer-Rao lower bound matrix of β and σ^2

$$\sum{}_{OLS} = \begin{bmatrix} \sigma^2 (X'X)^{-1} & 0 \\[2ex] 0 & \dfrac{2\sigma^4}{n-k} \end{bmatrix}$$

which means that the Cramer-Rao bound is attained for the covariance of b but not for s².

References

- Priestley, M.B. (1988) Non-linear and Non-stationary time series analysis, Academic Press. ISBN 0-12-564911-8

- Cramér, Harald (1946). Mathematical Methods of Statistics. Princeton, NJ: Princeton Univ. Press. ISBN 0-691-08004-6. OCLC 185436716

- Plackett, R.L. (1950). "Some Theorems in Least Squares". Biometrika. 37 (1–2): 149–157. JSTOR 2332158. MR 36980. doi:10.1093/biomet/37.1-2.149

- Rao, Calyampudi Radakrishna (1994). S. Das Gupta, ed. Selected Papers of C. R. Rao. New York: Wiley. ISBN 978-0-470-22091-7. OCLC 174244259

- Kay, S. M. (1993). Fundamentals of Statistical Signal Processing: Estimation Theory. Prentice Hall. p. 47. ISBN 0-13-042268-1

4

Standardized Coefficient and Statistical Hypothesis Testing

The estimates that arise due to the standardization of regression analysis, which cause the variables of dependent and independent variables to be 1. Unit normal scaling and unit length scaling are two common methods of standardized regression coefficients. The aspects elucidated in this chapter are of vital importance, and provide a better understanding of regression analysis and linear models.

Standardized Coefficient

In statistics, standardized coefficients or beta coefficients are the estimates resulting from a regression analysis that have been standardized so that the variances of dependent and independent variables are 1. Therefore, standardized coefficients refer to how many standard deviations a dependent variable will change, per standard deviation increase in the predictor variable. For univariate regression, the absolute value of the standardized coefficient equals the correlation coefficient. Standardization of the coefficient is usually done to answer the question of which of the independent variables have a greater effect on the dependent variable in a multiple regression analysis, when the variables are measured in different units of measurement (for example, income measured in dollars and family size measured in number of individuals).

Some statistical software packages like PSPP, SPSS and SYSTAT label the standardized regression coefficients as "Beta" while the unstandardized coefficients are labeled "B". Others, like DAP/SAS label them "Standardized Coefficient". Sometimes the unstandardized variables are also labeled as "b".

A regression carried out on original (unstandardized) variables produces unstandardized coefficients. A regression carried out on standardized variables produces standardized coefficients. Values for standardized and unstandardized coefficients can also be derived subsequent to either type of analysis.

Before solving a multiple regression problem, all variables (independent and dependent) can be standardized. Each variable can be standardized by subtracting its mean from each of its values and then dividing these new values by the standard deviation of the variable. Standardizing all variables in a multiple regression yields standardized regression coefficients that show the change in the dependent variable measured in standard deviations.

Advantages

       Standard coefficients' advocates note that the coefficients ignore the independent variable's scale of units, which makes comparisons easy.

Disadvantages

Critics voice concerns that such a standardization can be misleading. Since standardizing a variable removes the unit of measurement from its value, a standardized coefficient for a given relationship only represents its strength relative to the variation in the distributions. This invites bias due to sampling error when one standardizes variables using means and standard deviations based on small samples. Furthermore, a change of one standard deviation in one variable is only equivalent to a change of one standard deviation in another predictor insofar as the shapes of the two variables' distributions resemble one another. The meaning of a standard deviation may vary markedly between non-normal distributions (e.g., when skewed or otherwise asymmetrical). This underscores the importance of normality assumptions in parametric statistics, and poses an additional problem when interpreting standardized coefficient estimates that even nonparametric regression does not solve when dealing with non-normal distributions.

Standardized Regression Coefficients

Usually it is difficult to compare the regression coefficients because the magnitude of $\hat{\beta}_j$ reflects the units of measurement of jth explanatory variable X_j. For example, in the following fitted regression model

$$\hat{y} = 5 + X_1 + 1000X_2,$$

y is measured in liters, X_1 is milliliters and X_2 in liters. Although $\hat{\beta}_2 >> \hat{\beta}_1$ but effect of both explanatory variables is identical. One liter change in either X_1 and X_2 when other variable is held fixed produces the same change in $\hat{y}$.

Sometimes it is helpful to work with scaled explanatory and study variables that produces dimensionless regression coefficients.

These dimensionless regression coefficients are called as standardized regression coefficients.

There are two popular approaches for scaling which gives standardized regression coefficients.

We discuss them as follows:

Unit Normal Scaling

Employ unit normal scaling to each explanatory variable and study variable .

So define

$$Z_{ij} = \frac{x_{ij} - \bar{x}_j}{s_j}, \ i = 1, 2, ..., n, \ j = 1, 2, ..., k$$

$$y_i^* = \frac{y_i - \bar{y}}{s_y}$$

where $s_j^2 = \dfrac{1}{n-1}\sum_{i=1}^{n}(x_{ij}-\overline{x}_j)^2$ and $s_y^2 = \dfrac{1}{n-1}\sum_{i=1}^{n}(y_i-\overline{y})^2$

are the sample variances of jth explanatory variable and study variable, respectively.

All scaled explanatory variables and the scaled study variable have mean zero and sample variance unity, i.e., using these new variables, the regression model becomes

$$y_i^* = \gamma_1 z_{i1} + \gamma_2 z_{i2} + ... + \gamma_{ik} z_{ik} + \varepsilon, \; i = 1, 2, ..., n.$$

Such centering removes the intercept term from the model. The least squares estimate of $\gamma = (\gamma_1, \gamma_2, ..., \gamma_k)^{1}$, is

$$\hat{\gamma} = (Z'Z)^{-1} Z' y^*.$$

This scaling has a similarity to standardizing a normal random variable, i.e., observation minus its mean and divided by its standard deviation. So it is called as a unit normal scaling.

Unit Length Scaling

In unit length scaling, define

$$\omega_{ij} = \frac{x_{ij} - \overline{x}_j}{S_{jj}^{1/2}}, \; i = 1, 2, ..., n; \; j = 1, 2, ..., k$$

$$y_i^0 = \frac{y_i - \overline{y}}{SS_T^{1/2}}$$

Where

$$S_{jj} = \sum_{i=1}^{n}(x_{ij} - \overline{x}_j)^2$$

is the corrected sum of squares for j^{th} explanatory variables X_j and

$$S_T = SS_T = \sum_{i=1}^{n}(y_i - \overline{y})^2$$

is the total sum of squares.

In this scaling, each new explanatory variable W_j has mean $\overline{\omega}_j = \dfrac{1}{n}\sum_{i=1}^{n}\omega_{ij} = 0$ and length

$$\sqrt{\sum_{i=1}^{n}(\omega_{ij} - \overline{\omega}_j)} = 1.$$

In terms of these variables, the regression model is

$$y_i^0 = \delta_1 \omega_{i1} + \delta_2 \omega_{i2} + ... + \delta_k \omega_{ik} + \varepsilon_i, \ i = 1, 2, ..., n.$$

The least squares estimate of regression coefficient $\delta = (\delta_1, \delta_2, ..., \delta_k)'$ is

$$\overline{\delta} = (W'W)^{-1} W' y^0.$$

In such a case, the matrix $W'W$ is in the form of correlation matrix, i.e.,

$$W'W = \begin{pmatrix} 1 & r_{12} & r_{13} & \cdots & r_{1k} \\ r_{12} & 1 & r_{23} & \cdots & r_{2k} \\ r_{13} & r_{23} & 1 & \cdots & r_{3k} \\ \vdots & \vdots & \vdots & \ddots & \vdots \\ r_{1k} & r_{2k} & r_{3k} & \cdots & 1 \end{pmatrix}$$

Where

$$r_{ij} = \frac{\sum_{u=1}^{n} (x_{ui} - \overline{x}_i)(x_{uj} - \overline{x}_j)}{(S_{ii} S_{jj})^{1/2}}$$

$$= \frac{S_{ij}}{(S_{ii} S_{jj})^{1/2}}$$

is the simple correlation coefficient between the explanatory variables X_i and X_j.

Similarly,

$$W'y^0 = (r_{1y}, r_{2y}, ..., r_{ky})'$$

Where

$$r_{jy} = \frac{\sum_{u=1}^{n} (x_{uj} - \overline{x}_j)(y_u - \overline{y})}{(S_{jj} SS_T)^{1/2}} = \frac{S_{jy}}{(S_{jj} SS_T)^{1/2}}$$

is the simple correlation coefficient between j^{th} explanatory variable X_j and study variable y.

Note that it is customary to refer r_{ij} and r_{jy} as correlation coefficient though X_i 's are not random variable.

If unit normal scaling is used, then

$$Z'Z = (n-1)\, W'W.$$

So the estimates of regression coefficient in unit normal scaling (i.e., $\hat{\gamma}$), and unit length scaling (i.e., $\hat{\delta}$), are identical.

So it does not matter which scaling is used, this.

$$\hat{\gamma} = \hat{\delta}.$$

The regression coefficients obtained after such scaling, viz., $\hat{\gamma}$ or $\hat{\delta}$ are usually called standardized regression coefficients.

The relationship between the original and standardized regression coefficients is

$$b_j = \hat{\delta}_j \left(\frac{SS_T}{S_{jj}} \right)^{1/2}, j = 1, 2, ..., k$$

and

$$b_0 = \bar{y} - \sum_{j=1}^{k} b_j \bar{x}_j$$

where b_0 is the OLSE of intercept term and b_j are the OLSE of slope parameters β_j.

The Model in Deviation Form

The multiple linear regression model can also be expressed in the deviation form.

First all the data is expressed in terms of deviations from sample mean.

The estimation of regression parameters is performed in two steps:

First step: Estimate the slope parameters.

Second step: Estimate the intercept term.

The multiple linear regression model in deviation form is expressed as follows:

Let

$$A = I - \frac{1}{n} \ell \ell '$$

where $\ell = (1,1,...,1)'$ is a $n \times 1$ vector of each element unity. So

$$A = \begin{bmatrix} 1 & 0 & \cdots & 0 \\ 0 & 1 & \cdots & 0 \\ \vdots & \vdots & \ddots & \vdots \\ 0 & 0 & \cdots & 1 \end{bmatrix} - \frac{1}{n} \begin{bmatrix} 1 & 1 & \cdots & 1 \\ 1 & 1 & \cdots & 1 \\ \vdots & \vdots & \ddots & \vdots \\ 1 & 1 & \cdots & 1 \end{bmatrix}.$$

Then

$$\bar{y} = \frac{1}{n} \sum_{i=1}^{n} y_i = \frac{1}{n}(1,1,...,1) \begin{bmatrix} y_1 \\ y_2 \\ \vdots \\ y_n \end{bmatrix}$$

$$= \frac{1}{n} \ell' y$$

$$Ay = y - \ell\bar{y} = (y_1 - \bar{y}, y_2 - \bar{y}, ..., y_n - \bar{y})'.$$

Thus pre-multiplication of any column vector by A produces a vector showing those observations in deviation form.

Note that

$$A\ell = \ell - \frac{1}{n} \ell\ell'\ell$$

$$= \ell - \frac{1}{n} \ell.n$$

$$= \ell - \ell$$

$$= 0$$

and A is symmetric and idempotent matrix.

In the model

$$y = X\beta + \varepsilon,$$

the OLSE of β is

$$b = (X'X)^{-1} X'y$$

and residual vector is

$$e = y - Xb.$$

Note that $Ae = e$

Let the $n \times k$ matrix X is partitioned as

$$X = \begin{bmatrix} X_1 & X_2^* \end{bmatrix}$$

where $X_1 = (1,1,...,1)'$ is $n \times 1$ vector with all elements unity representing the intercept term, X_2^* is $n \times (k-1)$ matrix of observations of $(k-1)$ explanatory variables $X_2, X_3,..., X_k$ and OLSE $b = (b_1, b_2')$ is suitably partitioned with OLSE of intercept term β_1 as b_1 and b_2^* as a $(k-1) \times 1$ vector of OLSEs associated with $\beta_2, \beta_3,..., \beta_k$.

Then

$$y = X_1 b_1 + X_2^* b_2^* + e$$

Premultiplication by A gives

$$Ay = AX_1 b_1 + A X_2^* b_2^* + Ae$$
$$= A X_2^* b_2^* + e$$

Further, premultiplication by $X_2^{*'}$ gives

$$X_2^{*'} Ay = X_2^{*'} A X_2^* b_2^* + X_2^{*'} e$$
$$= X_2^{*'} A X_2^* b_2^*.$$

Since A is symmetric and idempotent, so

$$(A X_2^*)'(Ay) = (A X_2^*)'(A X_2^*) b_2^*.$$

This equation can be compared with the normal equations $X'y = X'Xb$ in the model $y = X\beta + \varepsilon$. Such a comparison yields the following conclusions:

- b_2^* is the sub-vector of OLSE.

- Ay is the study variables vector in deviation form.

- $A X_2^*$ is the explanatory variable matrix in deviation form.

- This is normal equation in terms of deviations. Its solution gives OLS of slope coefficients as

$$b_2^* = \left[(A X_2^*)'(A X_2^*) \right]^{-1} (A X_2^*)'(Ay).$$

The estimate of intercept term is obtained in the second step as follows:

Premultiplying $y = Xb + e$ by $\dfrac{1}{n}\ell'$ gives

$$\frac{1}{n}\ell'y = \frac{1}{n}\ell'Xb + \frac{1}{n}\ell'e$$

$$\bar{y} = \begin{bmatrix} 1 \, \bar{X}_2 \, \bar{X}_3 \ldots \bar{X}_k \end{bmatrix} \begin{pmatrix} b_1 \\ b_2 \\ \vdots \\ b_k \end{pmatrix} + 0$$

$$\Rightarrow b_1 = \bar{y} - b_2\bar{X}_2 - b_3\bar{X}_3 - \ldots - b_k\bar{X}_k.$$

Now we explain the various sums of squares in terms of this model.

The total sum of squares is

$$TSS = y'Ay.$$

Since

$$Ay = AX_2^*b_2^* + e$$
$$y'Ay = y'AX_2^*b_2^* + y'e$$
$$= (Xb + e)'AX_2^*b_2^* + y'e$$
$$= (X_1b_1 + X_2^*b_2^* + e)'AX_2^*b_2^* + (X_1b_1 + X_2^*b_2^* + e)'e$$
$$= b_2^{*'}X_2^{*'}AX_2^*b_2^* + e'e$$
$$TSS = SS_{reg} + SS_{res}$$

where the sum of squares due to regression is

$$SS_{res} = b_2^{*'}X_2^{*'}AX_2^*b_2^*$$

and the sum of squares due to residual is

$$SS_{res} = e'e.$$

Test of Hypothesis for $H_0 : R\beta = r$

We consider a general linear hypothesis that the parameters in β are contained in a subspace of parameter space for which $R\beta = r$ where R is a (J x k) matrix of known elements and r is a (J x 1) vector of known elements. Note that the matrix X'X is of full rank. In general, the null hypothesis

$$H_0 : R\beta = r$$

is termed as general linear hypothesis and

$$H_1 : R\beta \neq r$$

is the alternative hypothesis.

We assume that rank (R) = J, i.e., R is of full column rank, so that there is no linear dependence in the hypothesis.

Some special cases and interesting example of $H_0 : R\beta = r$ are as follows:

(i) $H_0 : R\beta_i = r$

Choose $J = 1$, $r = 0$, $R = [0,0,...,0,1,0,...,0]$ where 1 occurs at the ith position is R.

This particular hypothesis explains whether X_i has any effect on the linear model or not.

(ii) $H_0 : \beta_3 = \beta_4$ or $H_0 : \beta_3 - \beta_4 = 0$

Choose $J = 1$, $r = 0$, $R = [0,0,1,-1,0,...,0]$

(iii) $H_0 : \beta_3 = \beta_4 = \beta_5$

or $H_0 : \beta_3 - \beta_4 = 0$, $\beta_3 - \beta_5 = 0$

Choose $J = 1$, $r = (0,0)'$, $R = \begin{bmatrix} 0 & 0 & 1 & -1 & 0 & 0 & ... & 0 \\ 0 & 0 & 1 & 0 & -1 & 0 & ... & 0 \end{bmatrix}$.

(iv) $H_0 : \beta_3 + 5\beta_4 = 2$

Choose $J = 1$, $r = 0$, $R = [0, 0, 1, 5, 0,...,0]$

(v) $H_0 : \beta_2 = \beta_3 = ... = \beta_k = 0$

Choose

$$J = k - 1$$
$$r = (0, \quad 0,...,0)'$$
$$R = \begin{bmatrix} 0 & 1 & 0 & ... & 0 \\ 0 & 0 & 1 & ... & 0 \\ \vdots & \vdots & \vdots & \ddots & \vdots \\ 0 & 0 & 0 & ... & 1 \end{bmatrix}_{(k-1)\times k} = \begin{bmatrix} 0 \\ 0 \\ \vdots \\ 0 \end{bmatrix} \begin{matrix} I_{k-1} \end{matrix}.$$

This particular hypothesis explains the goodness of fit. It tells whether β_i has linear effect or not and are they of any importance. It also tests whether $X_2, X_3,..., X_k$ have any influence in the determination of y or not. Here $\beta_1 = 0$ is excluded because this involves additional implication that the mean level of y is zero. Our main concern is to know whether the explanatory variables help in explaining the variation in y around its mean value or not.

We develop the likelihood ratio test for $H_0 : R\beta = r$.

Likelihood Ratio Test

The likelihood ratio test statistic is

$$\lambda = \frac{\max L(\beta, \sigma^2 \mid y, X)}{\max L(\beta, \sigma^2 \mid y, X, R\,\beta = r)} = \frac{\hat{L}(\Omega)}{\hat{L}(\omega)}$$

where Ω denotes the whole parametric space and ω denotes the sample space.

If both the likelihoods are maximized, one constrained and the other unconstrained, then the value of the unconstrained will not be smaller than the value of the constrained. Hence $\lambda \geq 1$.

First we discus the likelihood ratio test for a simpler case when $R = 1_k$ and $r = \beta_0$, i.e., $\beta = \beta_0$.

This will give us better and detailed understanding for the minor details and then we generalize it for $R\,\beta = r$, in general.

Likelihood ratio test for $H_0 : \beta = \beta_0$

Let the null hypothesis related $k \times 1$ to vector β is $H_0 : \beta = \beta_0$

where β_0 is specified by the investigator. The elements of β_0 can take on any value, including zero.

The concerned alternative hypothesis is $H_1 : \beta \neq \beta_0$.

Since $\varepsilon \sim N(0, \sigma^2 I)$ in $y = X\beta + \varepsilon$, so $y \sim N(X\beta, \sigma^2 I)$. Thus the whole parametric space and sample space are Ω and ω respectively given by

$$\Omega : \left\{ (\beta, \sigma^2) : -\infty < \beta_i < \infty, \sigma^2 > 0, i = 1, 2, ..., k \right\}$$

$$\omega : \left\{ (\beta, \sigma^2) : \beta = \beta_0, \sigma^2 > 0 \right\}.$$

The unconstrained likelihood under Ω is

$$L(\beta, \sigma^2 \mid y, X) = \frac{1}{(2\pi\sigma^2)^{n/2}} \exp\left[-\frac{1}{2\sigma^2} (y - X\beta)'(y - X\beta) \right].$$

This is maximized over Ω when

$$\tilde{\beta} = (X'X)^{-1} X'y$$

$$\tilde{\sigma}^2 = \frac{1}{n}(y - X\tilde{\beta})'(y - X\tilde{\beta})$$

where $\tilde{\beta}$ and $\tilde{\sigma}^2$ are the maximum likelihood estimates of β and σ^2 which are the values obtained by maximizing the likelihood function. Thus

$$\hat{L}(\Omega) = \max L(\beta, \sigma^2 \mid y, X)$$

$$= \frac{1}{\left[\dfrac{2\pi}{n}(y - X\tilde{\beta})'(y - X\tilde{\beta})\right]^{\frac{n}{2}}} \exp\left[\frac{(y - X\tilde{\beta})'(y - X\tilde{\beta})}{\left(\dfrac{2(y - X\tilde{\beta})'(y - X\tilde{\beta})}{n}\right)}\right]$$

$$= \frac{n^{n/2}\exp\left(-\dfrac{n}{2}\right)}{(2\pi)^{n/2}\left[(y - X\tilde{\beta})'(y - X\tilde{\beta})\right]^{\frac{n}{2}}}.$$

The constrained likelihood under ω is

$$\hat{L}(\omega) = Max\, L(\beta, \sigma^2 \mid y, X, \beta = \beta_0)$$

$$= \frac{1}{(2\pi\sigma^2)^{n/2}} \exp\left[-\frac{1}{2\sigma^2}(y - X\beta_0)'(y - X\beta_0)\right].$$

Since β_0 is known, so the constrained likelihood function has an optimum variance estimator

$$\tilde{\sigma}_\omega^2 = \frac{1}{n}(y - X\beta_0)'(y - X\beta_0)$$

$$\hat{L}(\omega) = \frac{n^{n/2}\exp\left(-\dfrac{n}{2}\right)}{(2\pi)^{n/2}\left[(y - X\beta_0)'(y - X\beta_0)\right]^{n/2}}.$$

The likelihood ratio is

$$\frac{\hat{L}(\Omega)}{\hat{L}(\omega)} = \frac{\left(\dfrac{n^{n/2}\exp(-n/2)}{(2\pi)^{n/2}\left[(y - X\tilde{\beta})'(y - X\tilde{\beta})\right]^{n/2}}\right)}{\left(\dfrac{n^{n/2}\exp(-n/2)}{(2\pi)^{n/2}\left[(y - X\tilde{\beta}_0)'(y - X\tilde{\beta}_0)\right]^{n/2}}\right)}$$

$$= \left[\frac{(y - X\beta_0)'(y - X\beta_0)}{(y - X\tilde{\beta})'(y - X\tilde{\beta})}\right]^{n/2}$$

$$= \left(\frac{\tilde{\sigma}_\omega^2}{\tilde{\sigma}^2}\right)^{n/2}$$

$$= (\lambda)^{n/2}$$

where

$$\lambda = \frac{(y - X\beta_0)'(y - X\beta_0)}{(y - X\tilde{\beta})'(y - X\tilde{\beta})}$$

is the ratio of the quadratic forms. Now we simplify the numerator of λ as follows:

$$(y - X\beta_0)'(y - X\beta_0) = \left[(y - X\tilde{\beta}) + X(\tilde{\beta} - \beta_0)\right]' \left[(y - X\tilde{\beta}) + X(\tilde{\beta} - \beta_0)\right]$$

$$= (y - X\tilde{\beta})'(y - X\tilde{\beta}) + 2y'[I - X(X'X)^{-1}X']X(\tilde{\beta} - \beta_0) + (\tilde{\beta} - \beta_0)'X'X(\tilde{\beta} - \beta_0)$$

$$= (y - X\tilde{\beta})'(y - X\tilde{\beta}) + (\tilde{\beta} - \beta_0)'X'X(\tilde{\beta} - \beta_0).$$

Thus

$$\lambda = \frac{(y - X\tilde{\beta})'(y - X\tilde{\beta}) + (\tilde{\beta} - \beta_0)'X'X(\tilde{\beta} - \beta_0)}{(y - X\tilde{\beta})'(y - X\tilde{\beta})}$$

$$= 1 + \frac{(\tilde{\beta} - \beta_0)'X'X(\tilde{\beta} - \beta_0)}{(y - X\tilde{\beta})'(y - X\tilde{\beta})}$$

or $\lambda - 1 = \lambda_0 = \dfrac{(\tilde{\beta} - \beta_0)'X'X(\tilde{\beta} - \beta_0)}{(y - X\tilde{\beta})'(y - X\tilde{\beta})}$

where

$$0 \le \lambda_0 < \infty.$$

Distribution of Ratio of Quadratic Forms

Now we find the distribution of the quadratic forms involved in λ_0 to find the distribution of λ_0 as follows:

$$(y - X\tilde{\beta})'(y - X\tilde{\beta}) = \tilde{e}'\tilde{e}$$

$$= y'[I - X(X'X)^{-1}X']y$$

$$= y'\bar{H}y$$

$$= (X\beta + \varepsilon)'\bar{H}(X\beta + \varepsilon)$$

$$= \varepsilon'\bar{H}\varepsilon \quad (Using\ \bar{H}X = 0)$$

$$= (n - k)\hat{\sigma}^2.$$

Result: Let Z is a $n \times 1$ random vector that is distributed as $N(0, \sigma^2 I_n)$ and A is any symmetric idempotent $n \times n$ matrix of rank p then $\dfrac{Z'AZ}{\sigma^2} \sim \chi^2(p)$. Let B is another $n \times n$ symmetric idempotent matrix of rank q, then $\dfrac{Z'BZ}{\sigma^2} \sim \chi^2(q)$.

If AB = 0 then Z'AZ is distributed independently of Z'BZ.

So using this result, we have

$$\frac{y'\bar{H}y}{\sigma^2} = \frac{(n-k)\hat{\sigma}^2}{\sigma^2} \sim \chi^2(n\text{-}k).$$

Further, if H_0 is true, then $\beta = \beta_0$. Substituting $\beta = \beta_0$ in this expression, we have the quantity which is same as the numerator of λ_0. The numerator of λ_0 can be rewritten in a general form for any β as

$$(\tilde{\beta} - \beta)' X' X (\tilde{\beta} - \beta) = \varepsilon' X (X'X)^{-1} X' X (X'X)^{-1} X' \varepsilon$$

$$= \varepsilon' X (X'X)^{-1} X' \varepsilon$$

$$= \varepsilon' H \varepsilon$$

where H is an idempotent matrix with rank k.

Thus using this result, we have

$$\frac{\varepsilon' H \varepsilon}{\sigma^2} = \frac{\varepsilon' X' (X'X)^{-1} X' \varepsilon}{\sigma^2} \sim \chi^2(k).$$

Furthermore, the product of the quadratic form matrices in the numerator $\left(\varepsilon' \bar{H} \varepsilon \right)$ and in the denominator $\varepsilon' H \varepsilon$ of λ_0 is

$$\left[I - X(X'X)^{-1} X' \right] X(X'X)^{-1} X' = X(X'X)^{-1} X' - X(X'X)^{-1} X' X(X'X)^{-1} X' = 0$$

and hence the χ^2 random variables in numerator and denominator of λ_0 are independent. Dividing each of the χ^2 random variable by their respective degrees of freedom, we get

$$\lambda_1 = \left(\frac{\dfrac{(\tilde{\beta} - \beta_0)' X' X (\tilde{\beta} - \beta_0)}{\sigma^2}}{\dfrac{k}{\dfrac{(n-k)\hat{\sigma}^2}{\sigma^2}}} \right)$$

$$= \frac{(\tilde{\beta} - \beta_0)' X' X (\tilde{\beta} - \beta_0)}{k\hat{\sigma}^2}$$

$$= \frac{(y - X\beta_0)'(y - X\beta_0) - (y - X\tilde{\beta})'(y - X\tilde{\beta})}{k\hat{\sigma}^2}$$

$$\sim F(k, n-k) \text{ under } H_0.$$

Note that

$(y - X\beta_0)'(y - X\beta_0)$: Restricted error sum of squares

$(y - X\tilde{\beta})'(y - X\tilde{\beta})$: Unrestricted error sum of squares

Numerator of λ_1: Difference between the restricted and unrestricted error sum of squares.

The decision rule is to reject $H_0 : \beta = \beta_0$ at α level of significance whenever

$$\lambda_1 \geq F_\alpha(k, n-k)$$

where $F_\alpha(k, n-k)$ is the upper critical points on the central F-distribution with k and n - k degrees of freedom.

Likelihood ratio test for H₀ : Rβ = R

The same logic and reasons used in the development of likelihood ratio test for $H_0 : \beta = \beta_0$ can be extended to develop the likelihood ratio test for $H_0 : R\beta = R$ as follows:

$$\Omega = \left\{ (\beta, \sigma^2) : -\infty < \beta_i < \infty, \ \sigma^2 > 0, \ i = 1, 2, ..., k \right\}$$

$$\omega = \left\{ (\beta, \sigma^2) : -\infty < \beta_i < \infty, \ R\beta = r, \ \sigma^2 > 0 \right\}.$$

Let $\tilde{\beta} = (X'X)^{-1} X' y.$

Then

$$E(R\tilde{\beta}) = R\beta$$
$$V(R\tilde{\beta}) = E\left[R(\tilde{\beta} - \beta)(\tilde{\beta} - \beta)' R' \right]$$
$$= R V(\tilde{\beta}) R'$$
$$= \sigma^2 R(X'X)^{-1} R'.$$

Since $\tilde{\beta} \sim N\left[\beta, \sigma^2 (X'X)^{-1} \right]$

so $R\tilde{\beta} \sim N\left[R\beta, \sigma^2 R(X'X)^{-1} R' \right]$

$R\tilde{\beta} - r = R\tilde{\beta} - R\beta = R(\tilde{\beta} - \beta) \sim N\left[0, \sigma^2 R(X'X)^{-1} R' \right].$

There exists a matrix Q such that

$$\left[R(X'X)^{-1} R' \right]^{-1} = QQ'$$

and then

$$\xi = QR(b - \beta) \sim N(0, \sigma^2 I_n).$$

Therefore under $H_0 : R\beta - r = 0$,

$$\frac{\xi\xi'}{\sigma^2} = \frac{(R\tilde{\beta} - r)'QQ'(R\tilde{\beta} - r)}{\sigma^2}$$

$$= \frac{(R\tilde{\beta} - r)'[R(X'X)^{-1}R']^{-1}(R\tilde{\beta} - r)}{\sigma^2}$$

$$= \frac{(\tilde{\beta} - \beta)'R[R(X'X)^{-1}R']^{-1}R(\tilde{\beta} - \beta)}{\sigma^2}$$

$$= \frac{\varepsilon'X(X'X)^{-1}R'[R(X'X)^{-1}R']^{-1}R(X'X)^{-1}X'\varepsilon}{\sigma^2}$$

$$\sim \chi^2(J)$$

which is obtained as $X(X'X)^{-1}R'[R(X'X)^{-1}R']^{-1}R(X'X)^{-1}X'$ is an idempotent matrix and its trace is J which is the associated degrees of freedom with the respective quadratic form.

Also, irrespective of whether H_0 is true or not,

$$\frac{\tilde{e}'\tilde{e}}{\sigma^2} = \frac{(y - X\tilde{\beta})'(y - X\tilde{\beta})}{\sigma^2} = \frac{y'\bar{H}y}{\sigma^2} = \frac{(n-k)\hat{\sigma}^2}{\sigma^2} \sim \chi^2(n-k).$$

Moreover, the product of quadratic form matrices of $\tilde{e}'\tilde{e}$ and

$(\tilde{\beta} - \beta)'R'[R(X'X)^{-1}R']^{-1}R(\tilde{\beta} - \beta)$ is zero implying that both the quadratic forms are independent.

So in terms of likelihood ratio test statistic

$$\lambda_1 = \frac{\left(\dfrac{\dfrac{(R\tilde{\beta} - r)'[R(X'X)^{-1}R']^{-1}(R\tilde{\beta} - r)}{\sigma^2}}{J} \right)}{\dfrac{\left(\dfrac{(n-k)\hat{\sigma}^2}{\sigma^2} \right)}{n-k}}$$

$$= \frac{\left((R\tilde{\beta} - r)'[R(X'X)^{-1}R']^{-1}(R\tilde{\beta} - r) \right)}{J\hat{\sigma}^2}$$

$$\sim F(J, n-k) \text{ under } H_0.$$

So the decision rule is to reject H_o whenever

$$\lambda_1 \geq F_\alpha(J, n-k)$$

where $F_\alpha(J, n-k)$ is the upper critical points on the central F distribution with J and $(n - k)$ degrees of freedom.

Test of Significance of Regression (Analysis of Variance)

If we set $R = [0 \ \ I_{k-1}], r = 0$ then the hypothesis $H_0 : R\beta = R$ reduces to the following null hypothesis:

$$H_0 : \beta_2 = \beta_3 = \ldots = \beta_k = 0$$

which is tested against the alternative hypothesis

$H_1 : \beta_j \neq 0$ for at least one $j = 2, 3,\ldots, k$.

This hypothesis determines if there is a linear relationship between y and any set of the explanatory variables $X_2, X_3, \ldots, X_k$. Notice that X_1 corresponds to the intercept term in the model and hence $x_n = 1$ all $i = 1, 2,\ldots, n$.

This is an overall or global test of model adequacy. Rejection of the null hypothesis indicates that at least one of the explanatory variables among $X_2, X_3, \ldots, X_k$. contributes significantly to the model. This is called as analysis of variance.

Since $\varepsilon \sim N(0, \sigma^2 I)$,

so $y \sim N(X\beta, \sigma^2 I)$

$$b = (X'X)^{-1}X'y \sim N\left[\beta, \sigma^2(X'X)^{-1}\right].$$

Also

$$\hat{\sigma}^2 = \frac{SS_{res}}{n-k}$$

$$= \frac{(y-\hat{y})'(y-\hat{y})}{n-k}$$

$$= \frac{y'[I - X(X'X)^{-1}X']y}{n-k} = \frac{y'\bar{H}y}{n-k} = \frac{y'y - b'X'y}{n-k}.$$

Since $(X'X)^{-1}X'\bar{H} = 0$,

so b and are $\hat{\sigma}^2$ independently distributed.

Since $y'\bar{H}y = \varepsilon'\bar{H}\varepsilon$ and $\bar{H}$ is an idempotent matrix, so

$$\frac{SS_{res}}{\sigma^2} \sim \chi^2_{(n-k)},$$

i.e., central χ^2 distribution with (n - k) degrees of freedom.

Partition $X = [X_1, \quad X_2^*]$ where the submatrix X_2^* contains the explanatory variables $X_2, X_3, \ldots, X_k$ and

partition $\beta = [\beta_1, \quad \beta_2^*]$ where the subvector β_2^* contains the regression coefficients $\beta_2, \beta_3, \ldots, \beta_k$. Now partition the total sum of squares due to y's as

$$SS_T = y'Ay$$
$$= SS_{reg} + SS_{res}$$

Where

$$SS_{reg} = b_2^{*'}\,\mathrm{X}_2^{*'}\,A\,\mathrm{X}_2^*\,b_2^*$$

is the sum of squares due to regression and b_2^* is the OLSE of β_2^*.

The sum of squares due to residuals is given by

$$SS_{res} = (y - Xb)'(y - Xb)$$
$$= y'\bar{H}y$$
$$= SS_T - SS_{reg}.$$

Further

$\dfrac{SS_{reg}}{\sigma^2} \sim \chi^2_{(k-1)}\left(\dfrac{\beta_2^{*'}X_2^{*'}AX_2^*\beta_2^*}{2\sigma^2}\right)$, i.e., non-central χ^2 distribution with non – centrality parameter

$\dfrac{\beta_2^{*'}X_2^{*'}AX_2^*\beta_2^*}{2\sigma^2}$,

$\dfrac{SS_r}{\sigma^2} \sim \chi^2_{(n-1)}\left(\dfrac{\beta_2^{*'}X_2^{*'}AX_2^*\beta_2^*}{2\sigma^2}\right)$, , i.e., non-central χ^2 distribution with non – centrality parameter

$\dfrac{\beta_2^{*'}X_2^{*'}AX_2^*\beta_2^*}{2\sigma^2}$,

Since $X_2\bar{H} = 0$, so SS_{reg} and SS_{res} are independently distributed. The mean squares due to re-gression is

$$MS_{reg} = \dfrac{SS_{reg}}{k-1}$$

and the mean square due to error is

$$MS_{res} = \frac{SS_{res}}{n-k}.$$

Then

$$\frac{MS_{reg}}{MS_{res}} \sim F_{k-1,n-k}\left(\frac{\beta_2^{*'}X_2^{*'}AX_2^*\beta_2^*}{2\sigma^2}\right)$$

which is a non-central F -distribution with $(k-1)(n-k)$ degrees of freedom and non-centrality parameter $\dfrac{\beta_2^{*'}X_2^{*'}AX_2^*\beta_2^*}{2\sigma^2}$.

Under $H_0 : \beta_2 = \beta_3 = \ldots = \beta_k = 0,$

$$F = \frac{MS_{reg}}{MS_{res}} \sim F_{k-1,n-k}.$$

The decision rule is to reject at α level of significance whenever

$$F \geq F_\alpha(k-1, n-k)$$
.

The calculation of F-statistic can be summarized in the form of an analysis of variance (ANOVA) table given as follows:

Source of variation	Sum of squares	Degrees of freedom	Mean squares	F
Regression	SS_{reg}	$k-1$	$MS_{reg} = SS_{reg}/k-1$	F
Error	SS_{res}	$n-k$	$MS_{res} = SS_{res}/(n-k)$	
Total	SS_r	$n-1$		

Rejection of H_0 indicates that it is likely that atleast one $\beta_i \neq 0$ $(i = 1, 2, \ldots, k)$.

Test of Hypothesis on Individual Regression Coefficients

In case the test in analysis of variance is rejected, then another question arises is that which of the regression coefficients is/are responsible for the rejection of null hypothesis. The explanatory variables corresponding to such regression coefficients are important for the model.

Adding such explanatory variables also increases the variance of fitted values $\hat{y}$, so one needs to be cautious that only those regressors are added which are really important in explaining the response. Adding unimportant explanatory variables may increase the residual mean square which may decrease the usefulness of the model.

To test the null hypothesis $H_0 : \beta_j = 0$

versus the alternative hypothesis $H_1 : \beta_j \neq 0$

has already been discussed is the case of simple linear regression model. In present case, if H_0 is accepted, it implies that the explanatory variable X_j can be deleted from the model. The corresponding test statistic is

$$t = \frac{b_j}{se(b_j)} \sim t(\text{n}-\text{k}-1)\, under\, H_0$$

where the standard error of OLSE b_j of β_j is

$se(b_j) = \sqrt{\hat{\sigma}^2 C_{jj}}$ where C_{jj} denotes the jth diagonal element of $(X'X)^{-1}$ corresponding to b_j.

The decision rule is to reject H_0 at α level of significance if $|t| > t_{\frac{\alpha}{2}, n-k-1}$.

Note that this is only a partial or marginal test because b_j depends on all the other explanatory variables $X_i (i \neq j)$ that are in the model. This is a test of the contribution of X_j given the other explanatory variables in the model.

Statistical Hypothesis Testing

A statistical hypothesis, sometimes called confirmatory data analysis, is a hypothesis that is testable on the basis of observing a process that is modeled via a set of random variables. A statistical hypothesis test is a method of statistical inference. Commonly, two statistical data sets are compared, or a data set obtained by sampling is compared against a synthetic data set from an idealized model. A hypothesis is proposed for the statistical relationship between the two data sets, and this is compared as an alternative to an idealized null hypothesis that proposes no relationship between two data sets. The comparison is deemed *statistically significant* if the relationship between the data sets would be an unlikely realization of the null hypothesis according to a threshold probability—the significance level. Hypothesis tests are used in determining what outcomes of a study would lead to a rejection of the null hypothesis for a pre-specified level of significance. The process of distinguishing between the null hypothesis and the alternative hypothesis is aided by identifying two conceptual types of errors (type 1 & type 2), and by specifying parametric limits on e.g. how much type 1 error will be permitted.

An alternative framework for statistical hypothesis testing is to specify a set of statistical models, one for each candidate hypothesis, and then use model selection techniques to choose the most appropriate model. The most common selection techniques are based on either Akaike information criterion or Bayes factor.

Confirmatory data analysis can be contrasted with exploratory data analysis, which may not have pre-specified hypotheses.

Variations and Sub-classes

Statistical hypothesis testing is a key technique of both frequentist inference and Bayesian inference, although the two types of inference have notable differences. Statistical hypothesis tests define a procedure that controls (fixes) the probability of incorrectly *deciding* that a default position (null hypothesis) is incorrect. The procedure is based on how likely it would be for a set of observations to occur if the null hypothesis were true. Note that this probability of making an incorrect decision is *not* the probability that the null hypothesis is true, nor whether any specific alternative hypothesis is true. This contrasts with other possible techniques of decision theory in which the null and alternative hypothesis are treated on a more equal basis.

One naïve Bayesian approach to hypothesis testing is to base decisions on the posterior probability, but this fails when comparing point and continuous hypotheses. Other approaches to decision making, such as Bayesian decision theory, attempt to balance the consequences of incorrect decisions across all possibilities, rather than concentrating on a single null hypothesis. A number of other approaches to reaching a decision based on data are available via decision theory and optimal decisions, some of which have desirable properties. Hypothesis testing, though, is a dominant approach to data analysis in many fields of science. Extensions to the theory of hypothesis testing include the study of the power of tests, i.e. the probability of correctly rejecting the null hypothesis given that it is false. Such considerations can be used for the purpose of sample size determination prior to the collection of data.

The Testing Process

In the statistics literature, statistical hypothesis testing plays a fundamental role. The usual line of reasoning is as follows:

1. There is an initial research hypothesis of which the truth is unknown.

2. The first step is to state the relevant null and alternative hypotheses. This is important, as mis-stating the hypotheses will muddy the rest of the process.

3. The second step is to consider the statistical assumptions being made about the sample in doing the test; for example, assumptions about the statistical independence or about the form of the distributions of the observations. This is equally important as invalid assumptions will mean that the results of the test are invalid.

4. Decide which test is appropriate, and state the relevant test statistic T.

5. Derive the distribution of the test statistic under the null hypothesis from the assumptions. In standard cases this will be a well-known result. For example, the test statistic might follow a Student's t distribution or a normal distribution.

6. Select a significance level (α), a probability threshold below which the null hypothesis will be rejected. Common values are 5% and 1%.

7. The distribution of the test statistic under the null hypothesis partitions the possible values of T into those for which the null hypothesis is rejected—the so-called *critical region*—and those for which it is not. The probability of the critical region is α.

8. Compute from the observations the observed value t_{obs} of the test statistic T.

9. Decide to either reject the null hypothesis in favor of the alternative or not reject it. The decision rule is to reject the null hypothesis H_o if the observed value t_{obs} is in the critical region, and to accept or "fail to reject" the hypothesis otherwise.

An alternative process is commonly used:

1. Compute from the observations the observed value t_{obs} of the test statistic T.

2. Calculate the p-value. This is the probability, under the null hypothesis, of sampling a test statistic at least as extreme as that which was observed.

3. Reject the null hypothesis, in favor of the alternative hypothesis, if and only if the p-value is less than the significance level (the selected probability) threshold.

The two processes are equivalent. The former process was advantageous in the past when only tables of test statistics at common probability thresholds were available. It allowed a decision to be made without the calculation of a probability. It was adequate for classwork and for operational use, but it was deficient for reporting results.

The latter process relied on extensive tables or on computational support not always available. The explicit calculation of a probability is useful for reporting. The calculations are now trivially performed with appropriate software.

The difference in the two processes applied to the Radioactive suitcase example (below):

- "The Geiger-counter reading is 10. The limit is 9. Check the suitcase."

- "The Geiger-counter reading is high; 97% of safe suitcases have lower readings. The limit is 95%. Check the suitcase."

The former report is adequate, the latter gives a more detailed explanation of the data and the reason why the suitcase is being checked.

It is important to note the difference between accepting the null hypothesis and simply failing to reject it. The "fail to reject" terminology highlights the fact that the null hypothesis is assumed to be true from the start of the test; if there is a lack of evidence against it, it simply continues to be assumed true. The phrase "accept the null hypothesis" may suggest it has been proved simply because it has not been disproved, a logical fallacy known as the argument from ignorance. Unless a test with particularly high power is used, the idea of "accepting" the null hypothesis may be dangerous. Nonetheless the terminology is prevalent throughout statistics, where the meaning actually intended is well understood.

The processes described here are perfectly adequate for computation. They seriously neglect the design of experiments considerations.

It is particularly critical that appropriate sample sizes be estimated before conducting the experiment.

The phrase "test of significance" was coined by statistician Ronald Fisher.

Interpretation

If the p-value is less than the required significance level (equivalently, if the observed test statistic is in the critical region), then we say the null hypothesis is rejected at the given level of significance. Rejection of the null hypothesis is a conclusion. This is like a "guilty" verdict in a criminal trial: the evidence is sufficient to reject innocence, thus proving guilt. We might accept the alternative hypothesis (and the research hypothesis).

If the p-value is *not* less than the required significance level (equivalently, if the observed test statistic is outside the critical region), then the test has no result. The evidence is insufficient to support a conclusion. (This is like a jury that fails to reach a verdict.) The researcher typically gives extra consideration to those cases where the p-value is close to the significance level.

Some people find it helpful to think of the hypothesis testing framework as analogous to a mathematical proof by contradiction.

In the Lady tasting tea example (below), Fisher required the Lady to properly categorize all of the cups of tea to justify the conclusion that the result was unlikely to result from chance. He defined the critical region as that case alone. The region was defined by a probability (that the null hypothesis was correct) of less than 5%.

Whether rejection of the null hypothesis truly justifies acceptance of the research hypothesis depends on the structure of the hypotheses. Rejecting the hypothesis that a large paw print originated from a bear does not immediately prove the existence of Bigfoot. Hypothesis testing emphasizes the rejection, which is based on a probability, rather than the acceptance, which requires extra steps of logic.

"The probability of rejecting the null hypothesis is a function of five factors: whether the test is one- or two tailed, the level of significance, the standard deviation, the amount of deviation from the null hypothesis, and the number of observations." These factors are a source of criticism; factors under the control of the experimenter/analyst give the results an appearance of subjectivity.

Use and Importance

Statistics are helpful in analyzing most collections of data. This is equally true of hypothesis testing which can justify conclusions even when no scientific theory exists. In the Lady tasting tea example, it was "obvious" that no difference existed between (milk poured into tea) and (tea poured into milk). The data contradicted the "obvious".

Real world applications of hypothesis testing include:

- Testing whether more men than women suffer from nightmares

- Establishing authorship of documents

- Evaluating the effect of the full moon on behavior

- Determining the range at which a bat can detect an insect by echo

- Deciding whether hospital carpeting results in more infections

- Selecting the best means to stop smoking

- Checking whether bumper stickers reflect car owner behavior

- Testing the claims of handwriting analysts

Statistical hypothesis testing plays an important role in the whole of statistics and in statistical inference. For example, Lehmann (1992) in a review of the fundamental paper by Neyman and Pearson (1933) says: "Nevertheless, despite their shortcomings, the new paradigm formulated in the 1933 paper, and the many developments carried out within its framework continue to play a central role in both the theory and practice of statistics and can be expected to do so in the fore-seeable future".

Significance testing has been the favored statistical tool in some experimental social sciences (over 90% of articles in the *Journal of Applied Psychology* during the early 1990s). Other fields have favored the estimation of parameters (e.g., effect size). Significance testing is used as a sub-stitute for the traditional comparison of predicted value and experimental result at the core of the scientific method. When theory is only capable of predicting the sign of a relationship, a directional (one-sided) hypothesis test can be configured so that only a statistically significant result supports theory. This form of theory appraisal is the most heavily criticized application of hypothesis testing.

Cautions

"If the government required statistical procedures to carry warning labels like those on drugs, most inference methods would have long labels indeed." This caution applies to hypothesis tests and alternatives to them.

The successful hypothesis test is associated with a probability and a type-I error rate. The conclu-sion *might* be wrong.

The conclusion of the test is only as solid as the sample upon which it is based. The design of the experiment is critical. A number of unexpected effects have been observed including:

- The clever Hans effect. A horse appeared to be capable of doing simple arithmetic.

- The Hawthorne effect. Industrial workers were more productive in better illumination, and most productive in worse.

- The placebo effect. Pills with no medically active ingredients were remarkably effective.

A statistical analysis of misleading data produces misleading conclusions. The issue of data quality can be more subtle. In forecasting for example, there is no agreement on a measure of forecast accuracy. In the absence of a consensus measurement, no decision based on measurements will be without controversy.

The book *How to Lie with Statistics* is the most popular book on statistics ever published. It does not much consider hypothesis testing, but its cautions are applicable, including: Many claims are made on the basis of samples too small to convince. If a report does not mention sample size, be doubtful.

Hypothesis testing acts as a filter of statistical conclusions; only those results meeting a probability threshold are publishable. Economics also acts as a publication filter; only those results favorable to the author and funding source may be submitted for publication. The impact of filtering on publication is termed publication bias. A related problem is that of multiple testing (sometimes linked to data mining), in which a variety of tests for a variety of possible effects are applied to a single data set and only those yielding a significant result are reported. These are often dealt with by using multiplicity correction procedures that control the family wise error rate (FWER) or the false discovery rate (FDR).

Those making critical decisions based on the results of a hypothesis test are prudent to look at the details rather than the conclusion alone. In the physical sciences most results are fully accepted only when independently confirmed. The general advice concerning statistics is, "Figures never lie, but liars figure".

Examples

Lady Tasting Tea

In a famous example of hypothesis testing, known as the *Lady tasting tea*, Dr. Muriel Bristol, a female colleague of Fisher claimed to be able to tell whether the tea or the milk was added first to a cup. Fisher proposed to give her eight cups, four of each variety, in random order. One could then ask what the probability was for her getting the number she got correct, but just by chance. The null hypothesis was that the Lady had no such ability. The test statistic was a simple count of the number of successes in selecting the 4 cups. The critical region was the single case of 4 successes of 4 possible based on a conventional probability criterion ($< 5\%$; 1 of 70 $\approx$ 1.4%). Fisher asserted that no alternative hypothesis was (ever) required. The lady correctly identified every cup, which would be considered a statistically significant result.

Courtroom Trial

A statistical test procedure is comparable to a criminal trial; a defendant is considered not guilty as long as his or her guilt is not proven. The prosecutor tries to prove the guilt of the defendant. Only when there is enough charging evidence the defendant is convicted.

In the start of the procedure, there are two hypotheses H_0: "the defendant is not guilty", and H_1: "the defendant is guilty". The first one, H_0, is called the *null hypothesis*, and is for the time being accepted. The second one, H_1, is called the *alternative hypothesis*. It is the alternative hypothesis that one hopes to support.

The hypothesis of innocence is only rejected when an error is very unlikely, because one doesn't want to convict an innocent defendant. Such an error is called *error of the first kind* (i.e., the conviction of an innocent person), and the occurrence of this error is controlled to be rare. As a consequence of this asymmetric behaviour, the *error of the second kind* (acquitting a person who committed the crime), is often rather large.

	H₀ is true Truly not guilty	H₁ is true Truly guilty
Accept null hypothesis Acquittal	Right decision	Wrong decision Type II Error
Reject null hypothesis Conviction	Wrong decision Type I Error	Right decision

A criminal trial can be regarded as either or both of two decision processes: guilty vs not guilty or evidence vs a threshold ("beyond a reasonable doubt"). In one view, the defendant is judged; in the other view the performance of the prosecution (which bears the burden of proof) is judged. A hypothesis test can be regarded as either a judgment of a hypothesis or as a judgment of evidence.

Philosopher's Beans

The following example was produced by a philosopher describing scientific methods generations before hypothesis testing was formalized and popularized.

Few beans of this handful are white.

Most beans in this bag are white.

Therefore: Probably, these beans were taken from another bag.

This is an hypothetical inference.

The beans in the bag are the population. The handful are the sample. The null hypothesis is that the sample originated from the population. The criterion for rejecting the null-hypothesis is the "obvious" difference in appearance (an informal difference in the mean). The interesting result is that consideration of a real population and a real sample produced an imaginary bag. The philosopher was considering logic rather than probability. To be a real statistical hypothesis test, this example requires the formalities of a probability calculation and a comparison of that probability to a standard.

A simple generalization of the example considers a mixed bag of beans and a handful that contain either very few or very many white beans. The generalization considers both extremes. It requires more calculations and more comparisons to arrive at a formal answer, but the core philosophy is unchanged; If the composition of the handful is greatly different from that of the bag, then the sample probably originated from another bag. The original example is termed a one-sided or a one-tailed test while the generalization is termed a two-sided or two-tailed test.

The statement also relies on the inference that the sampling was random. If someone had been picking through the bag to find white beans, then it would explain why the handful had so many white beans, and also explain why the number of white beans in the bag was depleted (although the bag is probably intended to be assumed much larger than one's hand).

Clairvoyant Card Game

A person (the subject) is tested for clairvoyance. He is shown the reverse of a randomly chosen playing card 25 times and asked which of the four suits it belongs to. The number of hits, or correct answers, is called X.

As we try to find evidence of his clairvoyance, for the time being the null hypothesis is that the person is not clairvoyant. The alternative is, of course: the person is (more or less) clairvoyant.

If the null hypothesis is valid, the only thing the test person can do is guess. For every card, the probability (relative frequency) of any single suit appearing is 1/4. If the alternative is valid, the test subject will predict the suit correctly with probability greater than 1/4. We will call the probability of guessing correctly p. The hypotheses, then, are:

- null hypothesis $: H_0 : p = \frac{1}{4}$ (just guessing)

and

- alternative hypothesis $: H_1 : p > \frac{1}{4}$ (true clairvoyant).

When the test subject correctly predicts all 25 cards, we will consider him clairvoyant, and reject the null hypothesis. Thus also with 24 or 23 hits. With only 5 or 6 hits, on the other hand, there is no cause to consider him so. But what about 12 hits, or 17 hits? What is the critical number, c, of hits, at which point we consider the subject to be clairvoyant? How do we determine the critical value c? It is obvious that with the choice $c=25$ (i.e. we only accept clairvoyance when all cards are predicted correctly) we're more critical than with $c=10$. In the first case almost no test subjects will be recognized to be clairvoyant, in the second case, a certain number will pass the test. In practice, one decides how critical one will be. That is, one decides how often one accepts an error of the first kind – a false positive, or Type I error. With $c = 25$ the probability of such an error is:

$$P(reject\ H_0 \mid H_0\ is\ valid) = P(X = 25 \mid p = \tfrac{1}{4}) = \left(\tfrac{1}{4}\right)^{25} \approx 10^{-15},$$

and hence, very small. The probability of a false positive is the probability of randomly guessing correctly all 25 times.

Being less critical, with $c=10$, gives:

$$P(reject\ H_0 \mid H_0\ is\ valid) = P(X \geq 10 \mid p = \tfrac{1}{4}) = \sum_{k=10}^{25} P(X = k \mid p = \tfrac{1}{4}) \approx 0.07.$$

Thus, $c = 10$ yields a much greater probability of false positive.

Before the test is actually performed, the maximum acceptable probability of a Type I error (α) is determined. Typically, values in the range of 1% to 5% are selected. (If the maximum acceptable error rate is zero, an infinite number of correct guesses is required.) Depending on this Type 1 error rate, the critical value c is calculated. For example, if we select an error rate of 1%, c is calculated thus:

$$P(reject\ H_0 \mid H_0\ is\ valid) = P(X \geq c \mid p = \tfrac{1}{4}) \leq 0.01.$$

From all the numbers c, with this property, we choose the smallest, in order to minimize the probability of a Type II error, a false negative. For the above example, we select: $c = 13$.

Radioactive Suitcase

As an example, consider determining whether a suitcase contains some radioactive material. Placed under a Geiger counter, it produces 10 counts per minute. The null hypothesis is that no radioactive material is in the suitcase and that all measured counts are due to ambient radioactivity typical of the surrounding air and harmless objects. We can then calculate how likely it is that we would observe 10 counts per minute if the null hypothesis were true. If the null hypothesis predicts (say) on average 9 counts per minute, then according to the Poisson distribution typical for radioactive decay there is about 41% chance of recording 10 or more counts. Thus we can say that the suitcase is compatible with the null hypothesis (this does not guarantee that there is no radioactive material, just that we don't have enough evidence to suggest there is). On the other hand, if the null hypothesis predicts 3 counts per minute (for which the Poisson distribution predicts only 0.1% chance of recording 10 or more counts) then the suitcase is not compatible with the null hypothesis, and there are likely other factors responsible to produce the measurements.

The test does not directly assert the presence of radioactive material. A *successful* test asserts that the claim of no radioactive material present is unlikely given the reading (and therefore ...). The double negative (disproving the null hypothesis) of the method is confusing, but using a counter-example to disprove is standard mathematical practice. The attraction of the method is its practicality. We know (from experience) the expected range of counts with only ambient radioactivity present, so we can say that a measurement is *unusually* large. Statistics just formalizes the intuitive by using numbers instead of adjectives. We probably do not know the characteristics of the radioactive suitcases; We just assume that they produce larger readings.

To slightly formalize intuition: Radioactivity is suspected if the Geiger-count with the suitcase is among or exceeds the greatest (5% or 1%) of the Geiger-counts made with ambient radiation alone. This makes no assumptions about the distribution of counts. Many ambient radiation observations are required to obtain good probability estimates for rare events.

The test described here is more fully the null-hypothesis statistical significance test. The null hypothesis represents what we would believe by default, before seeing any evidence. Statistical significance is a possible finding of the test, declared when the observed sample is unlikely to have occurred by chance if the null hypothesis were true. The name of the test describes its formulation and its possible outcome. One characteristic of the test is its crisp decision: to reject or not reject the null hypothesis. A calculated value is compared to a threshold, which is determined from the tolerable risk of error.

Definition of Terms

The following definitions are mainly based on the exposition in the book by Lehmann and Romano:

Statistical hypothesis

> A statement about the parameters describing a population (not a sample).

Statistic

> A value calculated from a sample, often to summarize the sample for comparison purposes.

Simple hypothesis

> Any hypothesis which specifies the population distribution completely.

Composite hypothesis

Any hypothesis which does *not* specify the population distribution completely.

Null hypothesis (H_0)

A simple hypothesis associated with a contradiction to a theory one would like to prove.

Alternative hypothesis (H_1)

A hypothesis (often composite) associated with a theory one would like to prove.

Statistical test

A procedure whose inputs are samples and whose result is a hypothesis.

Region of acceptance

The set of values of the test statistic for which we fail to reject the null hypothesis.

Region of rejection / Critical region

The set of values of the test statistic for which the null hypothesis is rejected.

Critical value

The threshold value delimiting the regions of acceptance and rejection for the test statistic.

Power of a test ($1 - \beta$)

The test's probability of correctly rejecting the null hypothesis. The complement of the false negative rate, β. Power is termed sensitivity in biostatistics. ("This is a sensitive test. Because the result is negative, we can confidently say that the patient does not have the condition.")

Size

For simple hypotheses, this is the test's probability of *incorrectly* rejecting the null hypothesis. The false positive rate. For composite hypotheses this is the supremum of the probability of rejecting the null hypothesis over all cases covered by the null hypothesis. The complement of the false positive rate is termed specificity in biostatistics. ("This is a specific test. Because the result is positive, we can confidently say that the patient has the condition.")

Significance level of a test (α)

It is the upper bound imposed on the size of a test. Its value is chosen by the statistician prior to looking at the data or choosing any particular test to be used. It is the maximum exposure to erroneously rejecting H_0 he/she is ready to accept. Testing H_0 at significance level α means testing H_0 with a test whose size does not exceed α. In most cases, one uses tests whose size is equal to the significance level.

p-value

The probability, assuming the null hypothesis is true, of observing a result at least as extreme as the test statistic.

Statistical significance test

> A predecessor to the statistical hypothesis test. An experimental result was said to be statistically significant if a sample was sufficiently inconsistent with the (null) hypothesis. This was variously considered common sense, a pragmatic heuristic for identifying meaningful experimental results, a convention establishing a threshold of statistical evidence or a method for drawing conclusions from data. The statistical hypothesis test added mathematical rigor and philosophical consistency to the concept by making the alternative hypothesis explicit. The term is loosely used to describe the modern version which is now part of statistical hypothesis testing.

Conservative test

> A test is conservative if, when constructed for a given nominal significance level, the true probability of *incorrectly* rejecting the null hypothesis is never greater than the nominal level.

Exact test

> A test in which the significance level or critical value can be computed exactly, i.e., without any approximation. In some contexts this term is restricted to tests applied to categorical data and to permutation tests, in which computations are carried out by complete enumeration of all possible outcomes and their probabilities.

A statistical hypothesis test compares a test statistic (z or t for examples) to a threshold. The test statistic (the formula found in the table below) is based on optimality. For a fixed level of Type I error rate, use of these statistics minimizes Type II error rates (equivalent to maximizing power). The following terms describe tests in terms of such optimality:

Most powerful test

> For a given *size* or *significance level*, the test with the greatest power (probability of rejection) for a given value of the parameter(s) being tested, contained in the alternative hypothesis.

Uniformly most powerful test (UMP)

> A test with the greatest *power* for all values of the parameter(s) being tested, contained in the alternative hypothesis.

Common Test Statistics

One-sample tests are appropriate when a sample is being compared to the population from a hypothesis. The population characteristics are known from theory or are calculated from the population.

Two-sample tests are appropriate for comparing two samples, typically experimental and control samples from a scientifically controlled experiment.

Paired tests are appropriate for comparing two samples where it is impossible to control important variables. Rather than comparing two sets, members are paired between samples so the difference between the members becomes the sample. Typically the mean of the differences is then compared to zero. The common example scenario for when a paired difference test is appropriate is when a single set of test subjects has something applied to them and the test is intended to check for an effect.

Z-tests are appropriate for comparing means under stringent conditions regarding normality and a known standard deviation.

A *t*-test is appropriate for comparing means under relaxed conditions (less is assumed).

Tests of proportions are analogous to tests of means (the 50% proportion).

Chi-squared tests use the same calculations and the same probability distribution for different applications:

- Chi-squared tests for variance are used to determine whether a normal population has a specified variance. The null hypothesis is that it does.

- Chi-squared tests of independence are used for deciding whether two variables are associated or are independent. The variables are categorical rather than numeric. It can be used to decide whether left-handedness is correlated with libertarian politics (or not). The null hypothesis is that the variables are independent. The numbers used in the calculation are the observed and expected frequencies of occurrence (from contingency tables).

- Chi-squared goodness of fit tests are used to determine the adequacy of curves fit to data. The null hypothesis is that the curve fit is adequate. It is common to determine curve shapes to minimize the mean square error, so it is appropriate that the goodness-of-fit calculation sums the squared errors.

F-tests (analysis of variance, ANOVA) are commonly used when deciding whether groupings of data by category are meaningful. If the variance of test scores of the left-handed in a class is much smaller than the variance of the whole class, then it may be useful to study lefties as a group. The null hypothesis is that two variances are the same – so the proposed grouping is not meaningful.

In the table below, the symbols used are defined at the bottom of the table. Proofs exist that the test statistics are appropriate.

Name	Formula	Assumptions or notes
One-sample z-test	$$z = \frac{\bar{x} - \mu_0}{(\sigma / \sqrt{n})}$$	(Normal population or n > 30) and σ known. (z is the distance from the mean in relation to the standard deviation of the mean). For non-normal distributions it is possible to calculate a minimum proportion of a population that falls within k standard deviations for any k.
Two-sample z-test	$$z = \frac{(\bar{x}_1 - \bar{x}_2) - d_0}{\sqrt{\dfrac{\sigma_1^2}{n_1} + \dfrac{\sigma_2^2}{n_2}}}$$	Normal population and independent observations and σ_1 and σ_2 are known
One-sample t-test	$$t = \frac{\bar{x} - \mu_0}{(s / \sqrt{n})},$$ $$df = n - 1$$	(Normal population or n > 30) and σ unknown

Paired t-test	$$t = \frac{\bar{d} - d_0}{(s_d / \sqrt{n})},$$ $$df = n - 1$$	(Normal population of differences or n > 30) and σ unknown or small sample size $n < 30$
Two-sample pooled t-test, equal variances	$$t = \frac{(\bar{x}_1 - \bar{x}_2) - d_0}{s_p\sqrt{\dfrac{1}{n_1} + \dfrac{1}{n_2}}},$$ $$s_p^2 = \frac{(n_1 - 1)s_1^2 + (n_2 - 1)s_2^2}{n_1 + n_2 - 2},$$ $$df = n_1 + n_2 - 2$$	(Normal populations or $n_1 + n_2 > 40$) and independent observations and $\sigma_1 = \sigma_2$ unknown
Two-sample unpooled t-test, unequal variances (Welch's t-test)	$$t = \frac{(\bar{x}_1 - \bar{x}_2) - d_0}{\sqrt{\dfrac{s_1^2}{n_1} + \dfrac{s_2^2}{n_2}}},$$ $$df = \frac{\left(\dfrac{s_1^2}{n_1} + \dfrac{s_2^2}{n_2}\right)^2}{\dfrac{\left(\dfrac{s_1^2}{n_1}\right)^2}{n_1 - 1} + \dfrac{\left(\dfrac{s_2^2}{n_2}\right)^2}{n_2 - 1}}$$	(Normal populations or $n_1 + n_2 > 40$) and independent observations and $\sigma_1 \neq \sigma_2$ both unknown
One-proportion z-test	$$z = \frac{\hat{p} - p_0}{\sqrt{p_0(1 - p_0)}}\sqrt{n}$$	$n \cdot p_o > 10$ and $n(1 - p_o) > 10$ and it is a SRS (Simple Random Sample).
Two-proportion z-test, pooled for $H_0 : p_1 = p_2$	$$z = \frac{(\hat{p}_1 - \hat{p}_2)}{\sqrt{\hat{p}(1 - \hat{p})(\dfrac{1}{n_1} + \dfrac{1}{n_2})}}$$ $$\hat{p} = \frac{x_1 + x_2}{n_1 + n_2}$$	$n_1 p_1 > 5$ and $n_1(1 - p_1) > 5$ and $n_2 p_2 > 5$ and $n_2(1 - p_2) > 5$ and independent observations.
Two-proportion z-test, unpooled for $\lvert d_0 \rvert > 0$	$$z = \frac{(\hat{p}_1 - \hat{p}_2) - d_0}{\sqrt{\dfrac{\hat{p}_1(1 - \hat{p}_1)}{n_1} + \dfrac{\hat{p}_2(1 - \hat{p}_2)}{n_2}}}$$	$n_1 p_1 > 5$ and $n_1(1 - p_1) > 5$ and $n_2 p_2 > 5$ and $n_2(1 - p_2) > 5$ and independent observations.

Chi-squared test for variance	$$\chi^2 = (n-1)\frac{s^2}{\sigma_0^2}$$	Normal population
Chi-squared test for goodness of fit	$$\chi^2 = \sum^k \frac{(observed - expected)^2}{expected}$$	$df = k - 1 - \#\ parameters\ estimated$, and one of these must hold. • All expected counts are at least 5. • All expected counts are > 1 and no more than 20% of expected counts are less than 5
Two-sample F test for equality of variances	$$F = \frac{s_1^2}{s_2^2}$$	Normal populations Arrange so $s_1^2 \geq s_2^2$ and reject H_o for $$F > F(\alpha/2, n_1 - 1, n_2 - 1)$$
Regression t-test of $H_0 : R^2 = 0$.	$$t = \sqrt{\frac{R^2(n-k-1^*)}{1-R^2}}$$	Reject H_o for $t > t(\alpha/2, n-k-1^*)$ *Subtract 1 for intercept; k terms contain independent variables.

In general, the subscript 0 indicates a value taken from the null hypothesis, H_o, which should be used as much as possible in constructing its test statistic. ... *Definitions of other symbols:*

- α, the probability of Type I error (rejecting a null hypothesis when it is in fact true)

- n = sample size

- n_1 = sample 1 size

- n_2 = sample 2 size

- $\bar{x}$ = sample mean

- μ_0 = hypothesized population mean

- μ_1 = population 1 mean

- μ_2 = population 2 mean

- σ = population standard deviation

- σ^2 = population variance

- s = sample standard deviation

- $\sum^k$ = sum (of k numbers)

- s^2 = sample variance

- S_1 = sample 1 standard deviation

- S_2 = sample 2 standard deviation

- t = t statistic

- df = degrees of freedom

- $\bar{d}$ = sample mean of differences

- d_0 = hypothesized population mean difference

- S_d = standard deviation of differences

- χ^2 = Chi-squared statistic

- $\hat{p} = x/n$ = sample proportion, unless specified otherwise

- p_0 = hypothesized population proportion

- p_1 = proportion 1

- p_2 = proportion 2

- d_p = hypothesized difference in proportion

- $\min\{n_1, n_2\}$ = minimum of n_1 and n_2

- $x_1 = n_1 p_1$

- $x_2 = n_2 p_2$

- F = F statistic

Origins and Early Controversy

Significance testing is largely the product of Karl Pearson (p-value, Pearson's chi-squared test), William Sealy Gosset (Student's t-distribution), and Ronald Fisher ("null hypothesis", analysis of variance, "significance test"), while hypothesis testing was developed by Jerzy Neyman and Egon Pearson (son of Karl). Ronald Fisher began his life in statistics as a Bayesian (Zabell 1992), but Fisher soon grew disenchanted with the subjectivity involved (namely use of the principle of indifference when determining prior probabilities), and sought to provide a more "objective" approach to inductive inference.

Fisher was an agricultural statistician who emphasized rigorous experimental design and methods to extract a result from few samples assuming Gaussian distributions. Neyman (who teamed with the younger Pearson) emphasized mathematical rigor and methods to obtain more results from many samples and a wider range of distributions. Modern hypothesis testing is an inconsistent hybrid of the Fisher vs Neyman/Pearson formulation, methods and terminology developed in the early 20th century. While hypothesis testing was popularized early in the 20th century, evidence of its use can be found much earlier. In the 1770s Laplace considered the statistics of almost half a million births. The statistics showed an excess of boys compared to girls. He concluded by calculation of a p-value that the excess was a real, but unexplained, effect.

Fisher popularized the "significance test". He required a null-hypothesis (corresponding to a population frequency distribution) and a sample. His (now familiar) calculations determined whether to reject the null-hypothesis or not. Significance testing did not utilize an alternative hypothesis so there was no concept of a Type II error.

The p-value was devised as an informal, but objective, index meant to help a researcher determine (based on other knowledge) whether to modify future experiments or strengthen one's faith in the null hypothesis. Hypothesis testing (and Type I/II errors) was devised by Neyman and Pearson as a more objective alternative to Fisher's p-value, also meant to determine researcher behaviour, but without requiring any inductive inference by the researcher.

Neyman & Pearson considered a different problem (which they called "hypothesis testing"). They initially considered two simple hypotheses (both with frequency distributions). They calculated two probabilities and typically selected the hypothesis associated with the higher probability (the hypothesis more likely to have generated the sample). Their method always selected a hypothesis. It also allowed the calculation of both types of error probabilities.

Fisher and Neyman/Pearson clashed bitterly. Neyman/Pearson considered their formulation to be an improved generalization of significance testing.(The defining paper was abstract. Mathematicians have generalized and refined the theory for decades.) Fisher thought that it was not applicable to scientific research because often, during the course of the experiment, it is discovered that the initial assumptions about the null hypothesis are questionable due to unexpected sources of error. He believed that the use of rigid reject/accept decisions based on models formulated before data is collected was incompatible with this common scenario faced by scientists and attempts to apply this method to scientific research would lead to mass confusion.

The dispute between Fisher and Neyman–Pearson was waged on philosophical grounds, characterized by a philosopher as a dispute over the proper role of models in statistical inference.

Events intervened: Neyman accepted a position in the western hemisphere, breaking his partnership with Pearson and separating disputants (who had occupied the same building) by much of the planetary diameter. World War II provided an intermission in the debate. The dispute between

Fisher and Neyman terminated (unresolved after 27 years) with Fisher's death in 1962. Neyman wrote a well-regarded eulogy. Some of Neyman's later publications reported p-values and significance levels.

The modern version of hypothesis testing is a hybrid of the two approaches that resulted from confusion by writers of statistical textbooks (as predicted by Fisher) beginning in the 1940s. (But signal detection, for example, still uses the Neyman/Pearson formulation.) Great conceptual differences and many caveats in addition to those mentioned above were ignored. Neyman and Pearson provided the stronger terminology, the more rigorous mathematics and the more consistent philosophy, but the subject taught today in introductory statistics has more similarities with Fisher's method than theirs. This history explains the inconsistent terminology (example: the null hypothesis is never accepted, but there is a region of acceptance).

Sometime around 1940, in an apparent effort to provide researchers with a "non-controversial" way to have their cake and eat it too, the authors of statistical text books began anonymously combining these two strategies by using the p-value in place of the test statistic (or data) to test against the Neyman–Pearson "significance level". Thus, researchers were encouraged to infer the strength of their data against some null hypothesis using p-values, while also thinking they are retaining the post-data collection objectivity provided by hypothesis testing. It then became customary for the null hypothesis, which was originally some realistic research hypothesis, to be used almost solely as a strawman "nil" hypothesis (one where a treatment has no effect, regardless of the context).

A Comparison between Fisherian, Frequentist (Neyman–Pearson)

Fisher's null hypothesis testing	Neyman–Pearson decision theory
1. Set up a statistical null hypothesis. The null need not be a nil hypothesis (i.e., zero difference).	1. Set up two statistical hypotheses, H1 and H2, and decide about α, β, and sample size before the experiment, based on subjective cost-benefit considerations. These define a rejection region for each hypothesis.
2. Report the exact level of significance (e.g., p = 0.051 or p = 0.049). Do not use a conventional 5% level, and do not talk about accepting or rejecting hypotheses. If the result is "not significant", draw no conclusions and make no decisions, but suspend judgement until further data is available.	2. If the data falls into the rejection region of H1, accept H2; otherwise accept H1. Note that accepting a hypothesis does not mean that you believe in it, but only that you act as if it were true.
3. Use this procedure only if little is known about the problem at hand, and only to draw provisional conclusions in the context of an attempt to understand the experimental situation.	3. The usefulness of the procedure is limited among others to situations where you have a disjunction of hypotheses (e.g., either $\mu 1 = 8$ or $\mu 2 = 10$ is true) and where you can make meaningful cost-benefit trade-offs for choosing alpha and beta.

Early Choices of Null Hypothesis

Paul Meehl has argued that the epistemological importance of the choice of null hypothesis has gone largely unacknowledged. When the null hypothesis is predicted by theory, a more precise experiment will be a more severe test of the underlying theory. When the null hypothesis defaults to "no difference" or "no effect", a more precise experiment is a less severe test of the theory that motivated performing the experiment. An examination of the origins of the latter practice may therefore be useful:

1778: Pierre Laplace compares the birthrates of boys and girls in multiple European cities. He states: "it is natural to conclude that these possibilities are very nearly in the same ratio". Thus Laplace's null hypothesis that the birthrates of boys and girls should be equal given "conventional wisdom".

1900: Karl Pearson develops the chi squared test to determine "whether a given form of frequency curve will effectively describe the samples drawn from a given population." Thus the null hypothesis is that a population is described by some distribution predicted by theory. He uses as an example the numbers of five and sixes in the Weldon dice throw data.

1904: Karl Pearson develops the concept of "contingency" in order to determine whether outcomes are independent of a given categorical factor. Here the null hypothesis is by default that two things are unrelated (e.g. scar formation and death rates from smallpox). The null hypothesis in this case is no longer predicted by theory or conventional wisdom, but is instead the principle of indifference that lead Fisher and others to dismiss the use of "inverse probabilities".

Null Hypothesis Statistical Significance Testing

An example of Neyman–Pearson hypothesis testing can be made by a change to the radioactive suitcase example. If the "suitcase" is actually a shielded container for the transportation of radioactive material, then a test might be used to select among three hypotheses: no radioactive source present, one present, two (all) present. The test could be required for safety, with actions required in each case. The Neyman–Pearson lemma of hypothesis testing says that a good criterion for the selection of hypotheses is the ratio of their probabilities (a likelihood ratio). A simple method of solution is to select the hypothesis with the highest probability for the Geiger counts observed. The typical result matches intuition: few counts imply no source, many counts imply two sources and intermediate counts imply one source.

Neyman–Pearson theory can accommodate both prior probabilities and the costs of actions resulting from decisions. The former allows each test to consider the results of earlier tests (unlike Fisher's significance tests). The latter allows the consideration of economic issues (for example) as well as probabilities. A likelihood ratio remains a good criterion for selecting among hypotheses.

The two forms of hypothesis testing are based on different problem formulations. The original test is analogous to a true/false question; the Neyman–Pearson test is more like multiple choice. In the view of Tukey the former produces a conclusion on the basis of only strong evidence while the latter produces a decision on the basis of available evidence. While the two tests seem quite different both mathematically and philosophically, later developments lead to the opposite claim. Consider many tiny radioactive sources. The hypotheses become 0,1,2,3... grains of radioactive sand. There is little distinction between none or some radiation (Fisher) and 0 grains of radioactive sand versus all of the alternatives (Neyman–Pearson). The major Neyman–Pearson paper of 1933 also considered composite hypotheses (ones whose distribution includes an unknown parameter). An example proved the optimality of the (Student's) t-test, "there can be no better test for the hypothesis under consideration" (p 321). Neyman–Pearson theory was proving the optimality of Fisherian methods from its inception.

Fisher's significance testing has proven a popular flexible statistical tool in application with little mathematical growth potential. Neyman–Pearson hypothesis testing is claimed as a pillar of mathematical statistics, creating a new paradigm for the field. It also stimulated new applications in statistical process control, detection theory, decision theory and game theory. Both formulations have been successful, but the successes have been of a different character.

The dispute over formulations is unresolved. Science primarily uses Fisher's (slightly modified) formulation as taught in introductory statistics. Statisticians study Neyman–Pearson theory in graduate school. Mathematicians are proud of uniting the formulations. Philosophers consider them separately. Learned opinions deem the formulations variously competitive (Fisher vs Neyman), incompatible or complementary. The dispute has become more complex since Bayesian inference has achieved respectability.

The terminology is inconsistent. Hypothesis testing can mean any mixture of two formulations that both changed with time. Any discussion of significance testing vs hypothesis testing is doubly vulnerable to confusion.

Fisher thought that hypothesis testing was a useful strategy for performing industrial quality control, however, he strongly disagreed that hypothesis testing could be useful for scientists. Hypothesis testing provides a means of finding test statistics used in significance testing. The concept of power is useful in explaining the consequences of adjusting the significance level and is heavily used in sample size determination. The two methods remain philosophically distinct. They usually (but *not always*) produce the same mathematical answer. The preferred answer is context dependent. While the existing merger of Fisher and Neyman–Pearson theories has been heavily criticized, modifying the merger to achieve Bayesian goals has been considered.

Criticism

Criticism of statistical hypothesis testing fills volumes citing 300–400 primary references. Much of the criticism can be summarized by the following issues:

- The interpretation of a p-value is dependent upon stopping rule and definition of multiple comparison. The former often changes during the course of a study and the latter is unavoidably ambiguous. (i.e. "p values depend on both the (data) observed and on the other possible (data) that might have been observed but weren't").

- Confusion resulting (in part) from combining the methods of Fisher and Neyman–Pearson which are conceptually distinct.

- Emphasis on statistical significance to the exclusion of estimation and confirmation by repeated experiments.

- Rigidly requiring statistical significance as a criterion for publication, resulting in publication bias. Most of the criticism is indirect. Rather than being wrong, statistical hypothesis testing is misunderstood, overused and misused.

- When used to detect whether a difference exists between groups, a paradox arises. As improvements are made to experimental design (e.g., increased precision of measurement and sample size), the test becomes more lenient. Unless one accepts the absurd assumption that all sources of noise in the data cancel out completely, the chance of finding statistical significance in either direction approaches 100%.

- Layers of philosophical concerns. The probability of statistical significance is a function of decisions made by experimenters/analysts. If the decisions are based on convention they are termed arbitrary or mindless while those not so based may be termed subjective. To minimize type II errors, large samples are recommended. In psychology practically all null hypotheses are claimed to be false for sufficiently large samples so "...it is usually nonsensical to perform an experiment with the *sole* aim of rejecting the null hypothesis.". "Statistically significant findings are often misleading" in psychology. Statistical significance does not imply practical significance and correlation does not imply causation. Casting doubt on the null hypothesis is thus far from directly supporting the research hypothesis.

- "It does not tell us what we want to know".

Critics and supporters are largely in factual agreement regarding the characteristics of null hypothesis significance testing (NHST): While it can provide critical information, it is *inadequate as the sole tool for statistical analysis. Successfully rejecting the null hypothesis may offer no support for the research hypothesis.* The continuing controversy concerns the selection of the best statistical practices for the near-term future given the (often poor) existing practices. Critics would prefer to ban NHST completely, forcing a complete departure from those practices, while supporters suggest a less absolute change.

Controversy over significance testing, and its effects on publication bias in particular, has produced several results. The American Psychological Association has strengthened its statistical reporting requirements after review, medical journal publishers have recognized the obligation to publish some results that are not statistically significant to combat publication bias and a journal (*Journal of Articles in Support of the Null Hypothesis*) has been created to publish such results exclusively. Textbooks have added some cautions and increased coverage of the tools necessary to estimate the size of the sample required to produce significant results. Major organizations have not abandoned use of significance tests although some have discussed doing so.

Alternatives

The numerous criticisms of significance testing do not lead to a single alternative. A unifying position of critics is that statistics should not lead to a conclusion or a decision but to a probability or to an estimated value with a confidence interval rather than to an accept-reject decision regarding a particular hypothesis. It is unlikely that the controversy surrounding significance testing will be resolved in the near future. Its supposed flaws and unpopularity do not eliminate the need for an objective and transparent means of reaching conclusions regarding studies that produce statistical results. Critics have not unified around an alternative. Other forms of reporting confidence or uncertainty could probably grow in popularity. One strong critic of significance testing suggested a list of reporting alternatives: effect sizes for importance, prediction intervals for confidence, replications and extensions for replicability, meta-analyses for generality. None of these suggested alternatives produces a conclusion/decision. Lehmann said that hypothesis testing theory can be presented in terms of conclusions/decisions, probabilities, or confidence intervals. "The distinction between the ... approaches is largely one of reporting and interpretation."

On one "alternative" there is no disagreement: Fisher himself said, "In relation to the test of significance, we may say that a phenomenon is experimentally demonstrable when we know how to conduct an experiment which will rarely fail to give us a statistically significant result." Cohen, an influential critic of significance testing, concurred, "... don't look for a magic alternative to NHST *[null hypothesis significance testing]* ... It doesn't exist." "... given the problems of statistical induction, we must finally rely, as have the older sciences, on replication." The "alternative" to significance testing is repeated testing. The easiest way to decrease statistical uncertainty is by obtaining more data, whether by increased sample size or by repeated tests. Nickerson claimed to have never seen the publication of a literally replicated experiment in psychology. An indirect approach to replication is meta-analysis.

Bayesian inference is one proposed alternative to significance testing. (Nickerson cited 10 sources suggesting it, including Rozeboom (1960)). For example, Bayesian parameter estimation can provide rich information about the data from which researchers can draw inferences, while using uncertain priors that exert only minimal influence on the results when enough data is available. Psychologist John K. Kruschke has suggested Bayesian estimation as an alternative for the t-test. Alternatively two competing models/hypothesis can be compared using Bayes factors. Bayesian methods could be criticized for requiring information that is seldom available in the cases where significance testing is most heavily used. Neither the prior probabilities nor the probability distribution of the test statistic under the alternative hypothesis are often available in the social sciences.

Advocates of a Bayesian approach sometimes claim that the goal of a researcher is most often to objectively assess the probability that a hypothesis is true based on the data they have collected. Neither Fisher's significance testing, nor Neyman–Pearson hypothesis testing can provide this information, and do not claim to. The probability a hypothesis is true can only be derived from use of Bayes' Theorem, which was unsatisfactory to both the Fisher and Neyman–Pearson camps due to the explicit use of subjectivity in the form of the prior probability. Fisher's strategy is to sidestep this with the p-value (an objective *index* based on the data alone) followed by *inductive inference*, while Neyman–Pearson devised their approach of *inductive behaviour*.

Philosophy

Hypothesis testing and philosophy intersect. Inferential statistics, which includes hypothesis testing, is applied probability. Both probability and its application are intertwined with philosophy. Philosopher [[tyr]] wrote, "All knowledge degenerates into probability." Competing practical definitions of probability reflect philosophical differences. The most common application of hypothesis testing is in the scientific interpretation of experimental data, which is naturally studied by the philosophy of science.

Fisher and Neyman opposed the subjectivity of probability. Their views contributed to the objective definitions. The core of their historical disagreement was philosophical.

Many of the philosophical criticisms of hypothesis testing are discussed by statisticians in other contexts, particularly correlation does not imply causation and the design of experiments. Hypothesis testing is of continuing interest to philosophers.

Education

Statistics is increasingly being taught in schools with hypothesis testing being one of the elements taught. Many conclusions reported in the popular press (political opinion polls to medical studies) are based on statistics. An informed public should understand the limitations of statistical conclusions and many college fields of study require a course in statistics for the same reason. An introductory college statistics class places much emphasis on hypothesis testing – perhaps half of the course. Such fields as literature and divinity now include findings based on statistical analysis. An introductory statistics class teaches hypothesis testing as a cookbook process. Hypothesis testing is also taught at the postgraduate level. Statisticians learn how to create good statistical test procedures (like z, Student's t, F and chi-squared). Statistical hypothesis testing is considered a mature area within statistics, but a limited amount of development continues.

The cookbook method of teaching introductory statistics leaves no time for history, philosophy or controversy. Hypothesis testing has been taught as received unified method. Surveys showed that graduates of the class were filled with philosophical misconceptions (on all aspects of statistical inference) that persisted among instructors. While the problem was addressed more than a decade ago, and calls for educational reform continue, students still graduate from statistics classes holding fundamental misconceptions about hypothesis testing. Ideas for improving the teaching of hypothesis testing include encouraging students to search for statistical errors in published papers, teaching the history of statistics and emphasizing the controversy in a generally dry subject.

References

- Goldstein, H.; Healey, M.J.R. (1995). "The graphical presentation of a collection of means". Journal of the Royal Statistical Society. 158: 175–77. JSTOR view/2983411. doi:10.2307/2983411

- Larry D. Schroeder, David L. Sjoquist, Paula E. Stephan. (1986) Understanding regression analysis, Sage Publications. ISBN 0-8039-2758-4, p. 31-32

- Pratt, J. W. (1961). "Book Review: Testing Statistical Hypotheses. by E. L. Lehmann". Journal of the American Statistical Association. Taylor & Francis, Ltd. 56 (293): 163–167. JSTOR 2282344

- W. Härdle, M. Müller, S. Sperlich, A. Werwatz (2004), Nonparametric and Semiparametric Models, Springer, ISBN 3-540-20722-8

- Moore, David S. (1997). "New Pedagogy and New Content: The Case of Statistics". International Statistical Review. 65: 123–165. doi:10.2307/1403333

- Hinkelmann, Klaus and Kempthorne, Oscar (2008). Design and Analysis of Experiments. I and II (Second ed.). Wiley. ISBN 978-0-470-38551-7

- C. S. Peirce (August 1878). "Illustrations of the Logic of Science VI: Deduction, Induction, and Hypothesis". Popular Science Monthly. 13. Retrieved March 30, 2012

- Fisher, R (1955). "Statistical Methods and Scientific Induction" (PDF). Journal of the Royal Statistical Society, Series B. 17 (1): 69–78

- Stigler, Stephen M. (1986). The History of Statistics: The Measurement of Uncertainty before 1900. Cambridge, Mass: Belknap Press of Harvard University Press. p. 134. ISBN 0-674-40340-1

- Lehmann, E. L. (December 1993). "The Fisher, Neyman–Pearson Theories of Testing Hypotheses: One Theory or Two?". Journal of the American Statistical Association. 88 (424): 1242–1249. doi:10.1080/01621459.1993.10476404

- Morrison, Denton; Henkel, Ramon, eds. (2006) [1970]. The Significance Test Controversy. AldineTransaction. ISBN 0-202-30879-0

- Zabell, S (1989). "R. A. Fisher on the History of Inverse Probability". Statistical Science. 4 (3): 247–256. JSTOR 2245634. doi:10.1214/ss/1177012488

- Gigerenzer, G (November 2004). "Mindless statistics". The Journal of Socio-Economics. 33 (5): 587–606. doi:10.1016/j.socec.2004.09.033

- Harlow, Lisa Lavoie; Stanley A. Mulaik; James H. Steiger, eds. (1997). What If There Were No Significance Tests?. Lawrence Erlbaum Associates. ISBN 978-0-8058-2634-0

- Cornfield, Jerome (1976). "Recent Methodological Contributions to Clinical Trials" (PDF). American Journal of Epidemiology. 104 (4): 408–421

- Kline, Rex (2004). Beyond Significance Testing: Reforming Data Analysis Methods in Behavioral Research. Washington, D.C.: American Psychological Association. ISBN 9781591471189

- Neyman, J. (1937). "Outline of a Theory of Statistical Estimation Based on the Classical Theory of Probability". Philosophical Transactions of the Royal Society A. 236: 333–380. doi:10.1098/rsta.1937.0005

- Mayo, D. G.; Spanos, A. (2006). "Severe Testing as a Basic Concept in a Neyman–Pearson Philosophy of Induction". The British Journal for the Philosophy of Science. 57 (2): 323–357. doi:10.1093/bjps/axl003

Permissions

Index

A

Arithmetic Means, 91
Asymptotic Assumption, 50
Asymptotic Normality, 109, 111
Axis Regression Method, 54-55, 157

B

Bayesian Statistics, 12, 107, 145, 152
Biased Estimators, 184-185
Bienaymé Formula, 82, 84
Binomial Distribution, 8, 80

C

Centered Model, 103-105
Confidence Interval, 48-49, 122-127, 146, 148-149, 151, 229
Confidence Interval Estimation, 122, 124-125
Confidence Intervals, 21, 31, 48-49, 51, 120, 125, 134, 146, 151, 164, 229
Continuous Parameter Space, 115-116
Covariance, 11-12, 19, 35, 40, 42, 44, 71-72, 77, 81, 83-84, 86, 89, 92-100, 169, 176, 181, 183
Covariance Matrices, 97
Cramér-rao Bound, 184, 187, 189-190

D

Data Boundary Parameter-dependent, 111
Data Fitting, 26, 70, 75
Derived Linear Model, 134
Discrete Random Variable, 78, 80, 85, 96
Discrete Uniform Distribution, 115
Discrete Variables, 96

E

Environmental Science, 16
Estimation Methods, 5-6, 10, 41, 106
Explanatory Variable, 1, 41-44, 52, 101, 151, 164, 194-196, 211
Exponential Distribution, 28, 79
Extrapolation, 23, 166

F

Fair Die, 80
Financial Economics, 100
Finite Parameter Space, 115

Fisher Information, 28, 109, 112-114, 117, 184-185, 187, 189
Fitted Values, 179-180
Fixed-effects Models, 132
Full Rank, 67, 173, 178, 200
Functional Invariance, 113

G

Gauss-markov Theorem, 17, 30, 34-35, 69-70, 100, 173, 176-177
General Linear Models, 8

H

Heteroscedastic Models, 8
Hierarchical Linear Models, 9
Hypothesis Testing, 22, 31, 77, 127, 129-130, 135, 137, 193, 211-212, 214-217, 221, 225-231

I

Independent Measurements, 18
Intercept Term, 7, 46-47, 52, 104-105, 124-125, 141, 145, 171, 195, 197, 208
Interpolation, 23, 166

K

Known Mean, 146, 149, 189

L

Lasso Method, 36
Least Absolute Deviation Regression Method, 163
Least Squares, 2, 4-6, 8, 10-14, 38, 41, 44-45, 50, 53, 55-56, 58-60, 62, 66, 68-71, 73, 75-76, 100-101, 104-105, 122, 127, 168, 172-173, 175-176, 179, 181, 183, 192, 195-196
Least Squares Estimation, 6, 12, 53, 55
Least Squares Estimator, 70, 101, 173, 175-176, 179, 183
Least-squares Estimation, 10
Linear Algebra, 14, 99
Linear Combination, 4, 19, 31-32, 34-35, 83, 122, 125, 152, 175
Linear Least Squares, 13-14, 23, 27, 31-32, 36, 58, 60, 70, 73, 75-76

M

Matrix Notation, 22, 32, 64, 67, 83
Maximum-likelihood Estimation, 12, 107
Mixed-effects Models, 132

Molecular Biology, 99
Moment of Inertia, 92-93
Multiple Regression, 7-8, 21, 24, 41, 193
Multivariate Normal Distribution, 12, 178

N
Non-linear Least Squares, 23, 27, 31, 36, 58
Non-parametric Methods, 150
Normal Variance, 189
Normality Assumption, 48
Nuisance Parameters, 112

O
Ordinary Least Squares, 4, 10, 16, 20, 27, 36, 44-45, 50, 55-56, 58, 100-101, 172-173, 175, 183
Orthogonal Decomposition Methods, 67
Orthogonal Regression Method, 157

P
Parameter Confidence Limits, 72
Parameter Errors, 71-72
Poisson Distribution, 8, 79, 219
Potentially Relevant Variables, 40
Power Analysis, 138
Prediction Interval, 23, 44, 145-148, 150-153, 155
Predictive Bias, 154

Q
Quadratic Forms, 204, 207
Quadratic Model, 60

R
Random-effects Models, 132
Ratio Test, 201-202, 206-207
Regression Models, 2, 4, 8-9, 12-13, 15, 17, 23, 58, 75, 168

Residual Sum of Squares, 44, 100-102, 157, 182
Rounding Errors, 14, 76

S
Sample Covariance, 89, 95, 99
Sample Variance, 46, 86-89, 123, 148-149, 195, 224
Samuelson's Inequality, 91
Scalar Unbiased Case, 184
Semivariance, 93
Simple Regression Estimators, 51
Single-parameter Proof, 187
Slope Parameter, 52, 104, 122, 157
Spherical Errors, 178-179
Standardized Coefficient, 193-194
Statistical Assumptions, 18-20, 212
Statistically Dependent Variables, 84

T
Textbook Analysis, 133
Tikhonov Regularization, 36
Time Series Models, 108, 168
Trend Line, 15

U
Unbiasedness, 47, 185
Uncorrelatedness, 82, 98
Unit Length Scaling, 193, 195, 197
Unit-treatment Additivity, 133-135

W
Weighted Least Squares, 5, 8, 11, 19, 35
Weighted Linear Least Squares, 70
Weighted Sum, 35, 70, 84, 175